Praise for

"Always be recalculating. We get ——— —— —— lifetime. And that journey should be filled with recalculations. Those who live in fear of change will lose. Don't be afraid to recalculate. Lindsey Pollak is spot-on. Buy this book!"

—Kara Goldin, founder and CEO of Hint, Inc., and author of *Undaunted: Overcoming Doubts and Doubters*

"*Recalculating* is the career guide that I wish I'd had when I was going through my own career transition."

—Minda Harts, author of *The Memo: What Women of Color Need to Know to Secure a Seat at the Table*

"Lindsey Pollak is a master at career development with the experience, wisdom, and connections you are looking for to help you with your career. In *Recalculating*, Lindsey provides very specific and actionable advice on how to shift your mindset, set big goals, build your network, and get the job you want. This book should be *the* standard for anyone looking for a new job or career in 2020 and beyond. If you find yourself recalculating and searching for your next career, you've got to read this book! Start today!"

—Andy Storch, author of *Own Your Career Own Your Life: Stop Drifting and Take Control of Your Future*

"Living, working, and attempting to thrive during a pandemic mean all bets are off. Uncertainty is our common denominator. No one is better suited to be our guide through these challenging times than the bestselling author Lindsey Pollak. In *Recalculating*, she provides the master road map for anyone seeking a new role or new job, or taking on a career transition. It's a must-read."

—Susan McPherson, CEO and author of *The Lost Art of Connecting: The Gather, Ask, Do Method for Building Meaningful Business Relationships*

"Lindsey Pollak applies her time-tested workplace expertise to our current moment of uncertainty and deftly offers an inspiring and concrete path forward. *Recalculating* is chock-full of actionable advice for forging your dream career against the backdrop of a volatile job market, all filtered through Pollak's accessible and uplifting voice. That powerful combination makes *Recalculating* a must-read for early- and late-stage professionals alike, and a refreshingly bright moment of hope in an otherwise chaotic era."

—Jennifer Brown, founder and CEO of Jennifer Brown Consulting and author of *Inclusion: Diversity, the New Workplace & the Will to Change* and *How to be an Inclusive Leader: Your Role in Creating Cultures of Belonging Where Everyone Can Thrive*

"*Recalculating* is an amazing mindset-shift career-development book. During this digital age, social media has increased professionals' anxiety and depression as they navigate challenges such as unemployment and the COVID-19 pandemic. But this book provides innovative ways for professionals to adopt a growth mindset during difficult times. *Recalculating* will help professionals of all levels expand their professional networks, create work-life balance boundaries and negotiate their salaries. Our personal brands matter, but recalculating our mindsets produces unstoppable results."

— Kanika Tolver, CEO and author of *Career Rehab: Rebuild Your Personal Brand and Rethink the Way You Work*

"With *Recalculating*, Lindsey Pollak has provided a well-researched, inclusive, instructive, practical, and fun (!) read for folks figuring out how to build, grow, or restart their career in a post-COVID world. Her tips are specific. Her tactics are clear, and the benefits are immediately obvious. This is a must-read for career builders, shifters, and reinventors."

— Jeff Gothelf, author of *Forever Employable: How to Stop Looking for Work and Let Your Next Job Find You*

Recalculating

Recalculating

**NAVIGATE YOUR CAREER THROUGH
THE CHANGING WORLD OF WORK**

Lindsey Pollak

HARPER
BUSINESS

An Imprint of HarperCollinsPublishers

HarperCollins books may be purchased for educational, business,
or sales promotional use. For information, please email the Special
Markets Department at SPsales@harpercollins.com.

FIRST EDITION

Designed by Bonni Leon-Berman

Names: Pollak, Lindsey, author.
Title: Recalculating: navigate your career through the changing world
  of work / Lindsey Pollak.
Description: First edition. | New York, NY: Harper Business, [2021]
  | Includes bibliographical references and index. | Summary: "A
  leading workplace expert provides an inspirational, practical, and
  forward-looking career playbook for recent grads, career changers,
  and transitioning professionals looking to thrive in today's rapidly
  evolving workplace"—Provided by publisher.
Identifiers: LCCN 2020044792 (print) | LCCN 2020044793 (ebook)
  | ISBN 9780063067707 (trade paperback) | ISBN 9780063067714
  (ebook)
Subjects: LCSH: Vocational guidance—United States. | Career
  development—United States. | Job hunting—United States. | Career
  changes—United States.
Classification: LCC HF5382.5.U5 P56 2021 (print) | LCC
  HF5382.5.U5 (ebook) | DDC 650.1—dc23
LC record available at https://lccn.loc.gov/2020044792
LC ebook record available at https://lccn.loc.gov/2020044793

21 22 23 24 25   LSC   10 9 8 7 6 5 4 3 2 1

To Evan and Chloe,
and to every brave recalculator

Contents

Recalculating

Introduction

Do not judge me by my successes. Judge me by how many
times I fell down and got back up again.
—NELSON MANDELA

Do you remember exactly where you were when you first
heard about COVID-19?

I don't, and that seems strange to me. Usually when you
come across stories about previous history-altering moments—
JFK's assassination being the most famous—people always talk
about the exact moment they heard the news.

But the pandemic was different. In late 2019 to early 2020,
the virus was background news for the United States as it
spread through China and Europe. Then—as we all know too
well—by mid-March 2020, the pandemic had taken over
daily life, altering everyone's thoughts and actions and, most
tragically, taking far too many lives.

The impact of the pandemic on the economy and job market was immediate and devastating. Employers across regions
and industries announced cutbacks, furloughs, and layoffs.
Millions of employees who still had jobs were told to work

from home. Millions of others risked their lives to perform "essential" jobs.

And the pandemic was only one of several disruptive elements in 2020. According to government relations expert Bruce Mehlman, 2020 is the only year on record to include four "super-disruptors" to society: a recession, mass protests, an intense election, *and* a pandemic. To put things into perspective, only three other years since 1900 had even three.

Mehlman's July 2020 report was titled "The Great Acceleration: How 2020 Is Bringing the Future Faster," and that is exactly what the events of this historic year have done. COVID-19 wasn't the cause of all of the change it fostered; it was the colossal, global, unprecedented straw that broke the camel's back.

This is certainly true of the amplified attention the pandemic brought to our country's long struggles with racial injustice—a necessary reckoning that will be remembered as intensifying during the early months of the pandemic.

It is true of the spotlight the virus shone on the ongoing class divides in our country, the unequal access to quality healthcare and education, and the increased burden on women as caretakers of children and the elderly.

And it is true of career and workplace transitions as well. The pandemic accelerated shifts that were already well underway by 2020, including:

- increased automation, leading to the rise of certain jobs and industries and the fall of others
- an increasing combination of remote and in-person work

- increased action on issues of diversity, equity, and inclusion in the workplace
- more integration of "work" and "life"
- more concern about, and attention to, employee health and well-being
- more questioning of the value of higher education for career and financial success
- more Americans living and working into their seventies, eighties, and beyond

In October 2020, Erik Brynjolfsson, director of the Digital Economy Lab at the Stanford Institute for Human-Centered AI, remarked, "We've seen more changes in how we work over the past twenty weeks than we have over the past twenty years." Let's delve for a moment into the acceleration of remote work. According to the Federal Reserve, the share of the U.S. labor force working from home had already tripled in the time period from 2005 to the beginning of 2020. Then, within just the last few weeks of March, the numbers surged: as of March 15, 2020, 31 percent of U.S. workers had ever worked remotely. By April 2, that number had doubled, to 62 percent.

This meant that millions of professionals went from never having heard of Zoom to spending half their days on the videoconferencing app. Parents who had never worked from home were suddenly taking conference calls with toddlers on their laps or teenagers participating in remote high school classes in the same room. Leaders who had never talked openly about race were leading virtual town halls about implicit bias and microaggressions. Recruiters who normally touted the critical importance of a strong handshake were hiring employees without ever meeting them in person.

All of this change is merely referring to the experiences of those who were fortunate enough to retain their jobs. After 128 months of economic expansion in the United States—the longest in the history of our country—vast employment and business opportunities evaporated in a remarkably condensed period of time. Through just the first six months following the onset of the pandemic, one in four Americans put in new claims for unemployment benefits, twenty-one million Americans switched to part-time work because they couldn't find full-time jobs, and 55 percent of small businesses closed, never to reopen. In the same period, more than four million people graduated from college in hopes of starting their careers.

I know that I'm not reporting anything you don't already know. And words like "unprecedented," "historic," and "catastrophic" have become cliché when describing the events that have taken place since the emergence of COVID-19. But we cannot and should not gloss over the enormity of what has transpired in such a short period of time: since the beginning of 2020, every single person around the world has experienced a collective, unplanned, unwanted disruption—far worse for some than others—and it has changed every single one of our career trajectories in one way or another.

No matter what brought you personally to this moment, whether you are currently unemployed, employed, self-employed, or still not sure what day it is, I honor you for arriving here.

Living and working in these times is not for the faint of heart. *Recalculating* is here to help.

The Art of Recalculating

When I first started to think about all the people thrown unexpectedly into career transition during the initial stages of the pandemic, I kept imagining that moment when you're driving a car and the road forks, or you make a wrong turn, or you miss an exit. If you're using the GPS on your phone or another device, it will glitch for a few seconds and then a robotic voice will say, "Recalculating," sometimes over and over again (I admit I don't have the *best* sense of direction).

Now imagine every working adult on the planet in our cars, hearing this voice, all at the exact same time.

Maybe you're a recent grad trying to find your path after college. Maybe you were laid off from a longtime position in a "dying" industry and need to reinvent yourself in a new field. Maybe you're thinking about launching an entrepreneurial venture or freelance career or joining the "gig economy." Maybe the health crisis, the Black Lives Matter movement, or an environmental disaster has inspired you to pursue a different, more personally meaningful career path. Maybe you are a stay-at-home parent who wants to reenter the workforce. Maybe you've decided to recommit to a job that you had been losing interest in. Maybe you change your mind every hour and you're not sure exactly what you want from your career.

No matter what your specific situation, this book is for

you. Even if you are not currently in a moment of major transition, you are likely making readjustments and pivots all the time to keep up with the rapid pace of change. In the twenty-first-century workplace, standing still means being left behind. We are all recalculators now.

Recalculating is no doubt a challenge—but it's also an opportunity. As I started to think about what happens when a GPS is recalculating, I felt a growing sense of optimism. After all, when the navigation app is recalculating, it's demonstrating that there are multiple ways to get wherever you want to go. It factors in how far you've come already. And, if you decide to change your destination entirely, it can get you there as well.

When you recalculate, you open up infinite possibilities. As the Future Hunters, a futurist consultancy, affirmed in its report on the fateful year of 2020, "While the current pandemic is arguably the greatest global cataclysm since World War II, it also provides society with a seemingly unavoidable opportunity for pause, reset, and reimagination."

Throughout this book I'll share many powerful and inspiring stories of how an unexpected or unwanted career pivot led to greater satisfaction or success. You'll read stories from college students to senior executives to entrepreneurs. You'll read about my own experiences with recalculation. As you'll observe in these stories, sometimes the benefit of a detour was immediate and obvious to the person experiencing it; other times, the value of a recalculation took longer to feel like it paid off.

My conversations and research also revealed that recalculating isn't just a singular action at a particular moment in time, like deciding whether to turn right or left at a cross-roads. Rather, the most successful and happy professionals described themselves as frequent and deliberate recalculators. Some of their recalculations were big and bold; others were small and nuanced. But they treated recalculation as a vital skill in their professional toolkits—one that they applied over and over again to help guide them to success in the good times and the challenging ones.

Andy O'Hearn, who recalculated from a corporate career to a master's degree in library and information science, then back to a corporate career, put it this way: "Recalculation is not a phase; it's a mindset. Or, in more common parlance, it's not a bug; it's a feature."

Positive approaches to recalculation shouldn't have surprised me. I've been studying career and workplace issues for two decades and I have never met a single person who said, "Gosh, my career has just been a perfect path to success the whole way through!" Even the most prosperous, methodical, privileged people in the world have experienced failures, derailments, and disappointments on their career journeys. The people I admire most are the ones who remain agile and emerge stronger from every setback.

I'm not saying it's easy. There are plenty of catalysts for failure, derailment, and disappointment in one's career (sorry!), but the national and global events usually receive the most press. Think: Black Monday in 1987, 9/11 in 2001, the global financial crisis beginning in 2008, and now the multiple life-altering events of 2020. Economists say that catastrophic events like these, and the recessions that follow them, can permanently harm the employment and financial prospects of those who start their careers in such circumstances. It's tremendously challenging for people of other career stages and life situations, too.

I don't want to downplay this data and the very difficult task of making a living in a period of disruption and high unemployment. And I know that tough times are even tougher for those struggling with their own health or the health of a loved one, and disproportionally for BIPOC (Black, Indigenous, and people of color), people with different abilities, and members of other underrepresented groups. To cite the term created by civil rights advocate and scholar Kimberlé Crenshaw, people live intersectional lives. One's race, class, gender, sexual orientation, and other social categories "intersect" and overlap with one another, creating different experiences of discrimination or disadvantage, which are heightened in times of economic and social turmoil.

All of that being critically important to address and understand, I also refuse to write off anyone's career and economic success, even in today's rapidly changing, uncertain, and inequitable times. The goal of this book is to encourage you to take action on behalf of your personal career goals, and I will provide as many suggestions as possible on how to do so in tumultuous times. It is not easy and the system is often unfair, but while I advocate for systemic change, I will still encourage you on every page to take action toward your own success.

Inaction, in fact, is one of the biggest obstacles in the way of a successful recalculation. Consider this: Stephen Isherwood, CEO of the Institute of Student Employers in the U.K., told me during the first weeks of the pandemic that his biggest concern for the "Class of COVID-19" (a.k.a. 2020 grads) was that "they assume the labor market is dead and nobody is recruiting, so they just totally disengage." And then he shared a pretty astonishing statistic: during the global financial crisis, 50 percent of U.K. entry-level employers didn't fill all of their job vacancies. Why? Because students *weren't applying*. Multiple recruiters and university career services professionals told me the exact same thing when I asked for their top piece of advice for job seekers in challenging times: the biggest mistake by far is taking yourself out of the game because it seems impossible.

The same is true of the job market beyond entry-level hiring. According to research conducted by ZipRecruiter in July and August 2020, while unemployment jumped as a result of COVID-19, job search activity actually *declined* from pre-pandemic levels, as did Google searches for "how to find a job." While some people assumed the reason was that the U.S. government boosted unemployment benefits, ZipRecruiter's research found that "labor market conditions" were the main reason for job seekers' lack of action. People weren't applying because they falsely believed there were no jobs to apply for.

This is personal for me, because I made the same mistake in the early years of my own career. But I'll start my story by telling you where I am now: I currently have a successful business as a keynote speaker, corporate consultant, and author, advising organizations and individuals on career and workplace success. Over nearly twenty years, I've worked with more than 250 companies, law firms, universities, nonprofits, and government agencies to support everyone ranging from students to entry-level talent to senior leadership in navigating the ever-changing world of work.

I launched my own business in 2002 as a college campus

speaker, advising students on how to figure out their career paths. In 2007, I published my first book, *Getting from College to Career: 90 Things to Do Before You Join the Real World*, with that same goal in mind. For six years, I served as an official ambassador for LinkedIn, training over one hundred thousand students and job seekers on how to use the platform. During that time, my business pivoted to include keynote speaking engagements, along with providing career and management training to corporate employees at all levels.

In 2012, I published my second book, *Becoming the Boss: New Rules for the Next Generation of Leaders*, for the rising millennial workforce. In the years after that, I expanded my research again to focus on how different generations of employees can thrive together at work, offering new speeches and training programs on the topic and writing my third book in 2019, *The Remix: How to Lead and Succeed in the Multigenerational Workplace*.

I've been able to successfully recalculate my business several times, and I'm proud of what I've built. But my path, like most people's, has been full of ups and downs. I was first thwarted early in my career by an unprecedented event that was totally out of my control, and I did exactly what the experts warned against.

Back in 1999, I landed what was truly my entry-level dream job at a brand-new website called WorkingWoman.com. I had just earned my master's degree in women's studies and also wanted to experience the new world of digital media, so this job was ideal. I worked with smart, interesting people—mostly women, many of whom I'm still in touch with today—

and managed to catch the tail end of the dot-com bubble in New York City.

In April 2001, when I was twenty-six years old, the website ran out of funding and I was laid off. I knew it was coming, but I was still devastated. I always cite this layoff as the beginning of what turned out to be my entrepreneurship journey, but I didn't know that at the time. In the months after the layoff, I picked up some freelance writing work from a friend of a friend and a few very low-paid speaking engagements at local Rotary Clubs, chambers of commerce, and a few colleges, where I was invited to speak about my experience and offer advice for working in the new digital economy. But I was also applying for full-time jobs and wanted to continue building a corporate career.

And then came 9/11.

In the immediate aftermath of the terrorist attacks, hiring was at a complete standstill. I was very fortunate to have enough freelance work to pay my bills, but as the months progressed, I didn't do much to pursue a new job. I was intimidated by the unemployment reports and unsure about how to even reach out to companies in such uncertain and scary times. It felt selfish to worry about a job when my own city, and our country, had been attacked.

I continued to freelance and got some part-time work at a former colleague's marketing agency, and while I was grateful for the opportunities and income, I mostly just treaded water for the next year or so. To clarify, there is nothing wrong with freelancing or working in what is now called the gig economy. The issue was that I wanted a full-time job. I wanted to work

on-site with colleagues and learn the ropes of a big organization. But I was paralyzed. I felt disappointed and angry about
my layoff from a job I loved. As a result, I took myself out of
the game. I barely submitted any résumés, hardly spoke to
any professional contacts, and spent a lot of time feeling sorry
for myself. Thinking back on this experience, I'm ashamed
that I took no action despite having the privilege of being an
educated cisgender white woman with a supportive family
and good health. Maybe I wouldn't have been hired for any of
the jobs I didn't apply for, but maybe I would have. I'll never
know, and I regret it.

But now I know better. And, as Maya Angelou said, "Do
the best you can until you know better. Then when you know
better, do better."

Here is what I know now. No matter what stage of your
career you are in right now:

- You will experience bad economies.
- You will miss out on a job or client you really wanted.
- You will not get the role or promotion you expected.
- You will suffer through a terrible boss.
- You will have bad days.
- You will make mistakes, sometimes really big and painful
  ones.
- And, unfortunately, you may face discrimination.

But my greatest goal in writing this book is to ensure that
you will never, ever miss out on a career opportunity you
want because you didn't try for it. You can't recalculate if you
don't even start the journey.

The journey of this book will start, in Chapter One, with guidance on creating the best mindset for recalculation of any kind. In Chapter Two, we'll lay a strong foundation for current and future recalculations by considering career path shapes, self-assessment tools, education options, and time management strategies. Chapter Three will help you hone your career story and "marketing materials" that you will need, including advice on résumés, cover letters, LinkedIn profiles, and other social media. In Chapter Four, you'll explore the ins and outs of networking, both in person and virtually. Chapter Five dives deeply into every aspect of job hunting in today's times. Chapter Six offers guidance on turning any job into a positive experience and momentum-building opportunity. And Chapter Seven shares the stories of a wide variety of recalculators to inspire you on your path.

Feel free to jump around among the chapters and go right to the topics that are most urgent for you. At the same time, I also encourage you to read the sections that might seem *least* relevant to your own situation—sometimes that's where you'll gain the most creative ideas. I see this happen more than you might imagine: the strategies of an Uber driver inspire a business development idea in a corporate lawyer, the bravery of a college student changing majors galvanizes a decision by an entrepreneur, or the habits of a retiree motivate a frustrated job seeker.

When you embrace the art of recalculating, you accept the knowledge that your career path will not be a clear, straight line forward. You seek out change and challenge and discomfort, rather than avoiding it. This may sound scary, but it's a good thing! The sooner you learn how to handle false starts, detours, and disappointments—and even plan for them in advance—the smoother and more satisfying and more successful your journey will be. And sometimes you'll find that winding paths can take you in unexpected directions that are even better than you imagined.

Let's get started.

FIVE RULES FOR RECALCULATORS

I hereby welcome you as a member of the Recalculator Club. Here are our rules of the road, which I'll refer to throughout the book:

1. Embrace Creativity.

A successful recalculation requires you to try new things and break out of your comfort zone. You'll have to consider working in industries you may have disregarded (or had never heard of) before. You'll need to use your imagination to brainstorm new ways to describe your skills and qualifications. You'll be required to embrace new technologies and experiment with different methods of communication. You'll be asked to expand your network to include people who think, look, and work differently from you. You might even try TikTok!

If we've learned anything from the upheaval of 2020, it's that anything can happen and the world can change on a dime. Think of the doctors who found creative strategies to comfort and treat patients during the early days of the pandemic when they knew so little about the virus. Be inspired by the teachers who found creative ways to keep their students engaged when learning remotely. Now is not the time to be rigid. Be a little scrappy and relentlessly creative.

2. Prioritize Action.

Recalculation cannot take place only in your brain. You must, must, must take action. When you're in doubt, send an email.

When you're frustrated, call a friend or networking contact. When you're on the fence, apply for a job. Overthinking is an impediment to your success as a recalculator. In the words of Dale Carnegie, author of *How to Win Friends and Influence People*, "Inaction breeds doubt and fear. Action breeds confidence and courage. If you want to conquer fear, do not sit home and think about it. Go out and get busy."

3. Control What You Can.

I wish I could tell you how many résumés you'll need to send to land your dream job. I would love to tell you the exact date when job market conditions will be ideal for a career change. I dream about giving you the single salary negotiation tip that will guarantee you the highest income possible. Of course, all these wishes are impossible, because so much in life is out of our control.

The only sensible strategy is to put your energy into controlling what you can. You can control the amount of time you spend improving your résumé. You can control the number of people you connect and follow up with during your career change. You can control the amount of research you do about the appropriate salary range for your desired job, and you can control how much you rehearse your answer as to why you deserve even more. Control what you can and do your best to let the rest go.

4. Know Your Nonnegotiables.

Being creative or action-oriented, or letting some things go does *not* mean disregarding your values or morals or material needs. However, there is a possibility you'll have to make some trade-offs as part of your recalculation. You might take a step back in seniority, or stop making income for a few months, or work in an industry you never imagined, or move back in with your parents or a roommate. None of this is inherently "bad" or "good," "wrong" or "right." But it is your job to decide what compromises are okay to achieve your goals. (We'll spend more time on how to do this in Chapter Five.)

Some younger workers, for example, are most interested in paying back their student loans quickly and are willing to work long hours and sacrifice some social time to do so. Some caregivers will not accept a job that does not allow them to pick their child up at school or day care at the end of the day. Some people with medical conditions prioritize health insurance benefits above all else. Know your absolute musts and never waver.

5. Ask for Help.

You are never, ever alone in your current recalculation or at any point in your career journey. There are always people, organizations, websites, social media feeds, books, and articles that are here to help and support you (example: me!). All you have to do is ask. Don't know what salary request is appropriate in a particular industry? Ask. Don't know how to tie a tie for your job interview? Ask. Don't know what Slack is? Ask. You know when people say, "Don't hesitate to ask?" Don't. Ask early, ask often, ask forever.

Adjust Your Mindset

Approach Recalculation with a Positive Attitude— and a Little Less Instagram

> If you don't like something, change it. If you
> can't change it, change your attitude.
> —MAYA ANGELOU

My junior-year college roommate once showed up to our dorm room with one of those old-school, foldable trolley shopping carts. Instead of using a backpack like literally every other student on campus, she carted her books around in her little cart. When I questioned her about it (because, come on, who does that?), she said, "I decided it's cool."

This amazed me and my lifelong *but-what-will-everybody-think?* brain. I didn't know people were allowed to decide for themselves what was cool! But you can. And guess what: you are also allowed to decide for yourself how you are going to approach your recalculation.

Give it a try right now: Decide to believe that being a job

seeker or career changer or first-time freelancer in a time of unprecedented disruption is a good thing. Decide that you will make the most of it. Decide that it's your chance to shine. If you want, you can even decide that it's cool.

I don't mean to be flippant at all. Consciously choosing to own your mindset when approaching a tough situation is an essential ingredient in any successful career journey. I interviewed many recruiters, employers, and career experts for this book and asked each of them, "Where do you advise someone to begin their job search or entrepreneurial venture? What should they do on day one, minute one?" Every single expert recommended beginning in your own head.

Adunola Adeshola, a millennial career strategist and *Forbes* contributor, put it best: "You can have all the strategies in the world, but it won't help if you don't believe in yourself."

This is where we'll begin.

Tune Out the Noise

"Believe in yourself" might sound like a starry-eyed mantra, but there are tactical ways to implement this important advice. Here are four concrete steps to tuning out the noise that can drown your self-confidence:

Step 1: Tune Out Negative News.

The media is a major culprit of sowing self-doubt. News outlets love to generate negative narratives about the job market in particular. A two-second Internet search just yielded these confidence-bruising headlines:

- "4 Reasons the Job Search Is Really Hard"
- "The Future of Work for Millennials and Gen Z Is Bleak"
- "Not Getting Hired? Maybe People Think You're Too Old"
- "5 Reasons Why Employers Are Not Hiring Vets"
- "Why the Gig Economy Doesn't Really Work for Anyone"

I started to go down the rabbit hole of negative media headlines for additional examples to share with you, and then I realized it would take up the whole book. The point is: demoralizing articles about job hunting and business failures are everywhere and they will do nothing except discourage you. Don't read them, even when well-meaning friends and family forward them to you. Keep your focus on your own journey.

Speaking of well-meaning friends and family, sometimes (and I say this with love) you have to tune them out, too.

Step 2: Tune Out Unsupportive Loved Ones.

In my first book, *Getting from College to Career*, I wrote about "getting rid of the 'shoulds,'" which more often than not means giving less weight to the opinions of a relative, friend, or significant other about how you should manage your career. While your loved ones can and should support you in your career, sometimes their advice can be unhelpful and even harmful. Even with the best of intentions, they might project their own career desires (or regrets) onto you, or they might give you advice based on an outdated idea of what you want.

For example, when I was in third grade I wrote an essay about how I wanted to be a writer when I grew up. By the time I got to college, everyone seemed to be pushing me toward law school, which often happens to people who show some writing ability. My dad in particular was thrilled with the idea of Lindsey Pollak, Esq. I, however, abandoned all thoughts of becoming a lawyer about five minutes after a family friend invited me to visit her law firm and I learned more about what the job actually entails. (It's a great career, just not for me.)

Fast-forward many years later to the day I landed my first book deal. Ecstatic, I called my parents to share the news. "That's amazing!" my dad said with enthusiasm. "That will help you get into law school!"

Cue the facepalm emoji.

My experience is not uncommon. I've met people who have undergone years of schooling, pursued entire professions for decades, or taken over family businesses because they didn't want to disappoint their relatives. I know how hard it can be to feel that you might be letting down a loved one—especially if that person sacrificed a lot for you—but almost everyone I've met who pursued a career they never wanted has ended up feeling unfulfilled and switching career paths anyway. To make matters worse, they still had to share their true feelings with the person they were trying to please, at which point that person usually asked, "Why didn't you tell me sooner?" Honesty now can save a lot of pain and suffering later.

Step 3: Tune Out Anxiety-Inducing Social Media.

Putting aside for a moment the absurd amount of time many of us spend on social media, let's focus on the content we consume on apps like Instagram, Twitter, and Facebook. It's a wild understatement to say there is plenty of fodder for a negative mindset on social media. Research has found that passive scrolling through social media platforms undermines well-being, even if the scrolling only lasts for ten minutes a day.

So, let's nip the obvious stuff in the bud right now. If you know that a particular friend, celebrity, or influencer's feed is constantly negative, unattainably perfect, or undermines your confidence in any way, try muting or unfollowing it for a while and seeing if you feel more positive.

Step 4: Tune Out Negative Self-Talk.

The final tactical step to building more belief in yourself is the most difficult one: tuning out the noise in your own head. I've been trying to do this for most of my life and I still struggle, but I can confirm that even a small improvement in mindfulness is a game changer. The trick is not to totally clear your mind of all thoughts or be in a nonstop state of bliss, but to be a nonjudgmental observer of the thoughts passing by in your mind, "as if they are merely clouds in the sky," as my favorite meditation app likes to tell me.

For recalculators, these thought clouds might appear something like those negative media headlines:

- I'm too old to transition into the technology industry.
- No one will hire me because I'm on the autism spectrum.
- It's impossible to get an internship in fashion without family connections.
- Once that recruiter sees my GPA, I'll never get the job.

Could some of these thoughts be true? Yes. But are any of them definitely true? No. While you can't completely stop your mind from harboring these kinds of worries, you can do your best to minimize their impact on you. You can also stop these thoughts from migrating into your job search actions.

Here's an example of how this can happen and why it's a problem: One of the most common questions I receive when I lead public webinars for job seekers comes from older professionals concerned about ageism in the hiring process. To be totally clear, age discrimination is disturbingly prevalent and insidious, not to mention illegal when the discrimination is against a person who is forty or older (and I would argue that it should be illegal regardless of age). However, when I asked recruiters for their advice to older job seekers concerned about age discrimination, their answer surprised me.

While recruiters admit that ageism is a huge problem—in some industries, such as technology, more than others—they also tell me that, in some instances, it's actually the job candidate who draws attention to their own age. For instance, a candidate might start an interview by saying something like, "I know you probably think I'm too old for this position, but . . ." or "I'm probably old enough to be your mother, but . . ." or "You probably noticed I didn't list my college graduation year on my résumé, but . . ."

This is what it looks like to get in your own way.

How do you tone down the voice in your head so it doesn't derail you like this? The answer is to focus on what you *do* have to offer instead of emphasizing what you don't. This requires investigating the accuracy of your mind's internal dialogue.

One of Adunola Adeshola's clients, Sofia, wanted to transition from a marketing job in higher education to a marketing job in entertainment, her dream industry. Sofia started applying for positions she believed she was overqualified for in order to try to get her foot in the door. But even for these lower-level positions, employers rejected her for not having enough industry experience, or they just didn't respond to her application at all. She was understandably frustrated, disappointed, and down on herself.

The voice in Sofia's head told her she was deluded if she thought her experience in higher education could translate into a job in the entertainment industry. She would read job descriptions that looked amazing, then see one bullet point that didn't fit her skill set, and that would be enough for her to convince herself not to apply. Instead of letting thoughts of self-doubt pass her by, Sofia let herself wallow in the belief that she didn't have the right skills or experience to work in entertainment and she shouldn't even bother trying.

Adunola advised Sofia to fact-check that negative voice in her head that was sabotaging her. She encouraged Sofia to make new connections with a few professionals Sofia admired in the entertainment industry, asking for their candid assessments and advice. What Sofia learned was enlightening. It turned out the problem wasn't her experience or lack thereof; it was the language she was using to describe it.

For example, on her résumé, on her LinkedIn profile, and in her cover letters and interviews, Sofia talked about her work experiences with "students" and "professors." After chatting with people in the entertainment business, she changed her terminology, referring to these members of academia as "internal stakeholders" instead. She replaced every mention of "higher ed marketing experience" with simply "marketing experience." As a result of these small but important edits, the response from employers completely changed for the positive, which in turn boosted Sofia's confidence.

Sofia went from going unnoticed by employers to receiving multiple interviews in her dream industry. She was eventually hired at a television network where she still works today. When Sofia challenged her negative mindset of being rejected due to a lack of experience, she was able to find a relatively straightforward fix for her lack of job-searching success.

It's a hard reality to accept, but the voice in your own head isn't always on your side. If quieting your inner critic feels impossible, remember that we're not talking about silencing it entirely. Over the course of a job search, you'll sometimes feel discouraged and think pessimistic thoughts. That's normal. Just don't let those thoughts sink in too deeply. In the words of Martin Luther, "You cannot keep birds from flying over your head, but you can keep them from building a nest in your hair."

Quick Recalculation: What If I Need to Earn Money *Right Now*?

You might have picked up this book because, due to any number of reasons, you need to find a new source of income *fast*. Even if you are financially stable, it's never a bad idea to earn some extra cash and diversify your income streams.

For a recalculator, the ideal scenario is to pursue paid work in a field that is somehow related to the career you ultimately want to pursue. For example, a student named Juan at Texas A&M University–San Antonio wanted to be a doctor, and he needed to work his way through college. He was able to find an hourly virtual job as a medical scribe, taking notes during doctor-patient visits. This allowed Juan to pay the bills while having the additional benefit of immersing himself in the medical field and building relationships with doctors.

I know that it's not always possible to find a job related to your interests. In the months following the outbreak of COVID-19, many people took jobs they may not have considered even a few months earlier, such as working at the checkout of a grocery or retail store, telemarketing, delivering packages, or working in a warehouse or factory.

There is absolutely nothing wrong with taking a job "just for the money" at any point in your career. Recruiters consistently tell me that they are understanding of any decisions people make to support themselves and their families financially, especially in the wake of unanticipated circumstances like a pandemic, a natural disaster, a recession, or a personal issue like a health or family emergency.

Aside from pursuing full-time jobs to pay the bills, you might also consider a part-time or "gig" job to earn income during a transition. Many creative professionals took this route when Broadway productions shut down due to the pandemic. Matt Doyle, an actor in the musical *Company*, made money through multiple gigs, including singing personalized video messages on Cameo, teaching voice lessons on Zoom, gaming on Twitch, and writing comics on Webtoon. He even rented out his car through the car-sharing marketplace Turo. "It's hustle any way imaginable," he said in September 2020.

Even pre-pandemic, a 2018 Bankrate study found that 37 percent of Americans had a "side hustle" to make extra income, resulting in average earnings of $8,000 per year.

Here are some suggestions to consider for earning money while you make your way through this book:

- Offer your skills in home repair or landscaping (according to the Bankrate study, this is the top way Americans make side income).
- Drive for Lyft, Uber, or another ridesharing service.
- Sell your used clothing and accessories on a site like Poshmark or TheRealReal.
- Create arts or crafts to sell on Etsy.
- Market your skill in sales, writing, research, data entry, Web design, graphic design, translation, animation, social media, photo editing, or a variety of other areas on a freelance marketplace platform like Upwork, Freelancer, or Fiverr.
- Babysit or tutor children in your neighborhood. You can market yourself on a site like Care.com or Sittercity.
- Apply for a paid micro-internship project at ParkerDewey.com.

Use Envy as a Catalyst

It's absolutely critical to lower the volume on negativity, but there are also productive ways to use uncomfortable thoughts and emotions to support your career, especially in challenging times. Adunola advises her clients to do the following: "Look on LinkedIn to see people celebrate their jobs and accomplishments," she says. "Scroll through your feed and check out people who have gotten new jobs. Instead of angrily thinking, 'Why did they get that great job?' ask, 'How can I get a great job like that?' Think to yourself, 'That could be me!'"

Envy doesn't have to be a negative emotion—it can be a catalyst. Inspiration and extraordinarily detailed advice are a click away at all times. While passively scrolling through social media can be toxic and addictive, actively using it for research is a remarkably effective way to learn from successful people. (You don't even want to know how many hours I've spent poking around on the websites and social media feeds of other career advice authors and speakers.)

It's worthwhile to spend some time digging into the stories of people whose careers you admire. Position them in your mind as role models and inspirations. Study their career progressions. Explore their various social media feeds to see whom they follow, what content they post and share, and who else follows them. Read their blog posts. Listen to their podcasts.

It can be especially important for women, BIPOC, and members of underrepresented groups to find role models they can identify with. "There are almost always people who look like you who are thriving," says Adunola. "Start looking for evidence that what you want is possible and put your energy into thinking about how you can transition to being that person. It starts with thinking you are that person."

This was the case for Hannah White, currently student body vice president at the University of South Carolina. She first arrived on campus as a shy teenager, but when she saw that a Black man was student body president and one of her directors for freshman council, she remembers thinking, "I can do this. I want to be like that." Soon after, Hannah got involved in student government. "I am more willing to get involved when I see people like me in higher positions," she says.

If you're considering a career change, you can use Google and LinkedIn searches to research role models working within the industry or for a specific employer you're contemplating. If you're thinking of starting a business, you can listen to the start-up stories of successful entrepreneurs on the NPR podcast *How I Built This with Guy Raz* and check out the biographies and memoirs of successful contemporary and historical figures.

What exactly should you be reading and listening for? Seek out best practices that you can adopt, such as specific key words in their profiles or bios, skills they promote, or accomplishments they share. Find parallels within your own experience and skill set, and highlight those areas more during a job search or at your next performance review. Figure out which employers your role models have worked for; they might be potential prospects for you. Hunt for job titles you didn't know existed that might be a good fit for you. And don't forget to pay particular attention to that envy instinct. When a detail in someone's profile makes you especially jealous, that's a flashing neon sign that it's an important element of what you want for yourself.

Exercise: Hey Jealousy!

I was introduced to a version of this exercise by my first-ever life and career coach, Jennifer Macaluso-Gilmore. The instructions are to list ten people whose careers you admire in some way. Then fill out the questions provided for each person to turn your envy into a catalyst for action. (Bonus points if you recognize the Gen X song referenced in this exercise title . . .)

Name of person you envy	What you most envy about this person	Brainstorm one or more steps you can take to develop the same skills or qualities this person has
Carla Harris, author, speaker, singer, and managing partner at Morgan Stanley	She is the best public speaker I've ever witnessed	1. I can study videos of her speeches to observe her delivery, pacing, and storytelling and apply it to my own speeches. 2. I can reach out and verbalize my admiration for her, then ask if she can share some advice or assess a short video of my speaking if she is willing.
1.		

2.		
3.		
4.		
5.		
6.		

7.		
8.		
9.		
10.		

Adopt a Growth Mindset

It's impossible to address the topic of mindset without including the groundbreaking work of psychologist and Stanford University professor Carol S. Dweck, PhD, who literally wrote the book on the subject. In *Mindset: The New Psychology of Success*—a book that profoundly impacted me and that I recommend every chance I get—Dweck compares people with what she calls a "fixed" mindset to those with a "growth" mindset.

According to Dweck, people with a fixed mindset believe that they have a particular group of qualities, traits, and talents that are permanently set. A person with a fixed mindset believes statements like, "I'm just not artistic" or "I don't like change."

A person with a growth mindset takes the opposite approach, believing that almost any challenge can be overcome, any failure is an opportunity to grow, and any talent or ability can be improved through learning and action.

Dweck's research finds unequivocally that people with a growth mindset are happier and more successful. She writes, "For thirty years, my research has shown that *the view you adopt for yourself* profoundly affects the way you lead your life. It can determine whether you become the person you want to be and whether you accomplish the things you value."

That is a potent statement, especially in a time of global upheaval. And it seems the memo has been received: enrollments in growth mindset–related courses on the online learning marketplace Udemy surged more than 200 percent in the first few months of the pandemic.

If you lean toward a fixed mindset and are ready to try the growth approach, there is a very simple tool to use. It's a tiny little three-letter word that is incredibly powerful.

The word is: yet.

Here's how it works: whenever you find yourself in a fixed mindset during your recalculation—"I just don't have the experience start-ups want" or "I don't have the interview skills to succeed"—just add the word "yet" at the end of the sentence, and it will open up a whole new world of possibilities. Take a look:

"I just don't have the experience start-ups want."

End of discussion.

"I just don't have the experience start-ups want . . . yet."

Hmmm . . . I could research an online course that would give me some of the skills I saw in that start-up job description. Or I could volunteer a few hours a week for that new nonprofit in my town, which would have a similar environment to a start-up company. I'm going to try both and observe what happens.

"I don't have the interview skills to succeed."

It's hopeless.

"I don't have the interview skills to succeed . . . yet."

There are probably a thousand books on how to interview well. I'm going to borrow one from the library and read it for ten minutes every day before bed. I've also heard that I can use the services of my college career center no matter how long ago I graduated (Side note: this is true at almost all colleges and universities, along with many graduate schools. Check out Chapter Five for more advice on making the most of career centers.), so I'm going to send them an email asking for a mock interview session to get some tips. While I'm waiting for a reply, I'm just going to Google "how to do well at a job interview" and—oh wow, there are endless articles and videos! Why didn't I do this sooner?

The word "yet" is pretty remarkable for its power to open up possibilities. It doesn't guarantee that you can do absolutely anything ("I'm not a professional basketball player . . . yet" or "I haven't won an Oscar . . . yet") but it does open up the prospect that you can get a lot closer to any goal, or get a lot better at any skill, than you are today.

This reminds me of another book that changed my life: *The Artist's Way*, Julia Cameron's seminal work on overcoming blocks to creativity. (Embarrassing-to-admit fact: even though I had already cowritten a published book, I did not refer to myself as a "writer" until I read *The Artist's Way*.) At one point, Cameron specifically addresses people who feel they are too old to acquire a new skill—learning to play piano, for example:

Question: Do you know how old I will be by the time I learn to play the piano?

Answer: The same age you'll be if you don't.

It's never too late, or too early, to adopt a growth mindset and pursue a goal that is meaningful to you.

Exercise: Adopt a Growth Mindset

Developing a growth mindset requires not just a change in attitude but also a change in your daily habits. What takes place in your mindset must translate into action.

Complete this exercise for as many skills as you'd like to improve, especially for any perceived deficiencies you feel are holding you back from achieving your career goals.

1. Challenge a fixed mindset thought by adding the word "yet" to the end of it. _____

    _____

    _____

    _____

2. Brainstorm a variety of actions that can help you to learn, grow, or improve. _____

    _____

    _____

    _____

    _____

    _____

    _____

3. Pick the easiest or most appealing action to start with. Mark
   that action—or several elements of it—in your calendar to track
   your progress. _____

   _____

   _____

   _____

   _____

   _____

   _____

Let Go of What You're Leaving Behind

I have every confidence that your recalculation journey and
growth mindset will help lead you to a positive future, but I
also acknowledge that you may not have wanted to be in a
position to recalculate in the first place. Before taking action
to move forward with a positive attitude, you might need to
spend some time mourning what you're leaving behind.

Emma Lee Hartle, a student employment specialist at But-
ler County Community College in Pennsylvania, lives in Pitts-
burgh and was there in the 1970s and 1980s when all the steel
mills in the city closed down. She observed so many men who
spent their time just waiting for the mills to reopen, which
never happened. "When change is thrust upon you, do your
best to see it as an opportunity," she says. "I know a few of
these men who decided to follow their passion and became
chefs. They loved their new careers. Stay open to change and
embrace it as a new opportunity."

My youngest cousin, Olivia, was finishing her senior year at the University of California, Berkeley, when the pandemic cut short her college experience and postponed her graduation ceremony. While Olivia felt grateful to be healthy and have a safe home with her parents to return to when classes were canceled in March 2020, the lack of closure on her college experience was more upsetting than she had anticipated.

Even months after leaving campus, when she had secured a full-time job after having her first offer rescinded, Olivia told me, "My friends and I still talk all the time about the lack of an ending, not having a last week of classes, not walking across the stage for graduation. We are all in a weird place psychologically, like we can't acknowledge it's over. It's very sad. I had to undergo a pretty abrupt mindset change that my life is so much more serious now, working full-time and sitting in front of a computer twelve hours a day. I will never regain the experience I lost."

I have deep empathy for Olivia's experience. Life and career transitions are often extremely stressful, tiring, and anxiety-inducing. Honor that.

This is especially true if you are recalculating after a layoff. Layoffs are nothing to be ashamed of: 40 percent of Americans have been laid off or terminated from a job at least once (raising my hand here), and that statistic was published *before* COVID-19. Just because layoffs are common doesn't make them any easier, of course. Research shows that job loss can lead to serious mental and physical health issues. In some instances, it can take longer to get over the grief of a layoff than the death of a loved one.

Regardless of the reason for your current recalculation, it's important to pause and acknowledge any grief, anxiety, or other challenging emotions you might be feeling. And, if you are suffering, please seek professional help.

For advice on managing career-related grief or disappointment, I turned to two authorities on the topic: Zaheen Nanji, a resilience and leadership expert, and Jovian Zayne, a certified leadership and professional development coach. Here's what they had to say:

1. Reframe Your Situation.

It's perfectly normal and healthy to feel anxious or sad or any other emotion when you experience a layoff, business closure, or other career disappointment. One way to process those emotions is to consciously focus on the positive opportunities in any situation, or what Zaheen refers to as "reframing."

For example, it's natural after a layoff to worry about paying your rent or mortgage or to feel anxious about landing a new job. But it doesn't help to dwell on those feelings. To reframe, Zaheen recommends that you pause and say to yourself, "What can I find that is positive about this situation?" For example, maybe you had been thinking about seeking a new job anyway, or getting a certification you've always wanted to have. Make a conscious choice to change your perspective and focus on the opportunities in your situation rather than the dangers. Zaheen advises that you continually ask yourself, "What do I want now?" to move your attention from what you've lost to what you want to create.

This was the experience of Robin Solow, who was laid off

in April 2020 because of COVID-19, after having held various HR positions at a large luxury retailer for thirteen years. "When I really stopped to think about it, I wasn't happy in the company for a long time," she shared. "I had felt stuck, like there wasn't any place for me to go there. I don't know where the feeling came from, but from the minute I was let go I somehow knew it was an opportunity." As you'll read more about in Chapter Four, Robin ultimately ended up networking her way to a job at a commercial real estate start-up. "I never would have considered working at a start-up before the pandemic," she told me. "Suddenly, nothing felt risky anymore. Instead, it was exciting."

2. Practice Gratitude.

"When we find ourselves upset or complaining about things we can't control, those moments are the perfect invitation to turn to gratitude. I encourage people in transition to write down five things you are unequivocally grateful for," advises Jovian. "The positive emotion of gratitude doesn't just come because you say you want to be grateful. You have to invite it in and really see it for what it is."

Oprah Winfrey is a huge proponent of gratitude lists as a daily practice that can change your life. She has said, "I know for sure that appreciating whatever shows up for you in life changes your personal vibration. You radiate and generate more goodness for yourself when you're aware of all you have and not focusing on your have-nots."

Try this right now: jot down five things in your life that you're grateful for and see if it improves your mood.

3. Rely on Your Support System.

Another recommendation from Jovian is to write down five ways other people have served you or supported you in any way. This will help you remember that there are people operating on your behalf and that you are not living in isolation and solitude. Do your best to carefully curate the people who surround you and focus on those who are a positive influence. Reach out to them for support, camaraderie, and encouragement. Remember recalculator rule #5: ask for help.

4. Be of Service.

The corollary of asking for help is offering help. "The act of helping someone else helps you feel better, too," says Zaheen. To affirm your value to others, Jovian recommends making a list of five ways you have felt good about supporting someone else, perhaps through volunteer work, helping a colleague, coordinating a child's birthday party, or listening to the concerns of a friend. If you think you have nothing to give, you will not be in the right place to find your next opportunity.

Robin Solow put this advice into practice while she was job hunting. She took on a virtual temp job screening people who signed up to be contact tracers for the state of New York in the early months of the pandemic. "It wasn't as much about the money as about having a purpose," Robin says. As a bonus, when she was interviewing for the job she ultimately landed, she told the CEO about the temp role and he was impressed that she had chosen to give back in such a challenging time.

5. Set and Accomplish Short-Term Goals.

To avoid the feeling of overwhelm during a stressful time, Zaheen recommends setting small goals that help you to build a sense of accomplishment and positive momentum. For example, Zaheen and her husband own a wellness center in Alberta, Canada, which was forced to close when COVID-19 hit. In the same month as the shutdown orders, Zaheen's mother-in-law passed away and her father-in-law moved into their home. "We felt like our world was out of control," she acknowledges.

While Zaheen and her husband did the difficult and emotional work of moving as much of their business online as possible and cleaning out his parents' home of forty years, they also set small goals for what they could achieve to accomplish some "small wins": fixing a broken door at the wellness center and repainting a wall. "When the future feels uncertain," she says, invoking recalculator rule #3, "it helps to focus on what you can control."

Prioritize Self-Care

The process of career transition is rarely quick or easy, so self-care and patience with oneself are essential components of a successful recalculation over the long term. And since there is no one way that works for everyone to de-stress, relax, or recharge, try experimenting a bit to find what helps you stay positive and calm. Approach this task with your growth mindset: consider self-care another skill to improve on, little by little, over time. Any strategies you discover now will help keep you strong and focused during future transitions as well.

I'm particularly concerned about the mental health of younger millennials and Gen Zs, who are in the early years of their careers in today's challenging times. One 2020 study found that nearly two-thirds of millennials and Gen Zs said they feel anxiety nearly every day. That's *triple* the rate of baby boomers.

As someone who struggles with anxiety myself (and wishes I had acknowledged and treated it earlier in my life), I've had many heart-to-heart conversations over the years with younger professionals who are experiencing mental health issues. One such person is Kristen Frohlich, who approached me at one of my speaking events in New York City, where she was part of an audience of interns from a large advertising conglomerate. We immediately bonded because we had both been college dorm RAs, and when Kristen followed up with me on LinkedIn a year later to share some great job news with me, she also shared that she had struggled with anxiety during her college years.

"I think there's a lot of pressure on college students to know *exactly* what they want to do when they finish school," she told me. "This causes students a lot of anxiety—I know it did for me—about the uncertainty of the future that lies ahead of graduation."

As a result of this pressure, Kristen completed seven (!) internships between her first and third years at The College of New Jersey (TCNJ). Additionally, during her senior year she took on more than a full course load in order to graduate a semester early, all while being an RA, the vice president of her coed professional business fraternity, and the president of another honor society. It was just too much to juggle. The responsibilities and stress she felt affected her so much that her chest began to hurt from stress, and she eventually had a panic attack that landed her in the ER.

"From that moment on, I knew I had to cut back and start committing to less so I could focus on myself and my mental health," Kristen says. She started taking an online meditation course, and also took a stress management class offered by TCNJ the following semester to help manage her anxiety. Kristen's self-care routine allowed her to take a step back and evaluate the direction she wanted to take after college. She decided to focus on a single internship at a marketing agency, where she now works full-time while continuing to make self-care a priority.

Exercise: Build a Self-Care Practice

To support your self-care efforts, here are a wide variety of practices to test out and keep in your toolkit during your recalculation and whenever you are feeling overwhelmed throughout your career journey. I encourage you to try an action from this list that you've never considered before:

❑ Take five deep breaths
❑ Take a nap or go to bed an hour earlier than usual
❑ Exercise
❑ Take a movement class like yoga, dance, Pilates, or martial arts
❑ Listen to music

- ❑ Play a video game
- ❑ Work on a crossword puzzle, word search, or Sudoku
- ❑ Watch a TV show or movie
- ❑ Play with a pet
- ❑ Pursue a hobby you loved as a kid, like building LEGO or doing a jigsaw puzzle
- ❑ Garden, birdwatch, hike, fly a kite, or just spend time in nature
- ❑ Meditate (apps like Calm and Headspace provide guided meditations)
- ❑ Read a book
- ❑ Call a friend
- ❑ Cook or bake
- ❑ Get or give yourself a manicure or pedicure
- ❑ Visit a museum or tour one online
- ❑ Draw, color, or paint
- ❑ Play solitaire on the computer or with an actual deck of cards
- ❑ Pray or attend a religious service
- ❑ Other: _____
  _____
  _____

During a job search or career transition, it can be very tempting to want to spend every minute of your day Seriously Working on Your Career, but that is simply unsustainable. To set yourself up for success, consider asking a friend to check in on your progress once a week or keep a simple log of your self-care actions. What works for me is to build accountability around self-care: my friend Ilana and I send a quick text to each other when we finish our daily meditations.

I want to be totally clear that managing your mindset and taking time for self-care are absolutely, positively fundamental components of a productive recalculation. You can make some progress without paying them any mind, but eventually you'll be running on fumes.

Stay in Your Lane

Let's address one final mindset adjustment to support your recalculation. You can and should adopt a growth mindset to improve any skill, talent, or quality you want to build or enhance on your career journey. And self-care will help you maintain your energy and positivity. But there are some facts that you just can't change, such as your age, your gender, your race, your level of education at this moment, and the current economy and unemployment rate.

Keep this in mind if you find yourself starting to wonder who else might be applying for the same jobs, business loans, accelerator programs, or graduate programs you are, or if you start ruminating on questions like, "Are the other candidates more qualified? Do they speak another language? Do they have higher test scores? Is their employment history less patchy than mine?"

Whenever your thoughts wander into this territory, STOP and remember recalculator rule #3: control what you can. There is no upside to thinking about your competition because the answer to all of the above questions is: it doesn't matter, because there is absolutely nothing you can do about it. Stay in your own lane.

For some closing advice on staying in one's own lane, I reached out to a literal and figurative expert. Lamarr Pottinger is a former star collegiate track-and-field athlete at Eastern Illinois University, where he was conference champion three out of his four years in the 60-meter high hurdles. Today, he is the associate director of leadership development for the National Collegiate Athletic Association (NCAA), where he manages several leadership programs for student athletes transitioning into their careers. When I asked his advice on recalculating, he told me that one of the biggest challenges for nonathletes (ahem, like me) is to become as accustomed to failure as elite athletes are.

Athletes, Lamarr explained, are conditioned to consider failures as opportunities to improve. Case in point: in a single season, Babe Ruth broke the major league baseball record for both the most home runs *and* the most strikeouts. "Every strike brings me closer to the next home run," he said, in a textbook display of the growth mindset.

Lamarr encourages student athletes transitioning into post-college careers—and all recalculators—to condition themselves similarly to view career-related failures as opportunities to improve. How do you go about doing that? Think like a hurdler.

"To use a track analogy," Lamarr says, "you have to focus on your lane. You can't look to the side of you. You can't compare yourself to the person next to you. You have to focus on what you want to do in the race. If you don't focus forward and you are looking to the side, especially as a hurdler, you are going to fall and you won't get over those hurdles."

When you're on hurdle one, you can't look ahead to hurdle ten or you'll fall. In your career, that means focusing on the next deadline you have to meet, the next cover letter you have to write, the next networking call you've set up. Don't worry about what will happen if you don't get an offer from the company you'll be interviewing with this afternoon; just focus on preparing as much as you can for the interview itself. Don't wonder what other job seekers wrote in their cover letters; just focus on writing an authentic letter of your own.

And when you do stumble and fall or knock down a hurdle—which you will, because we all do—get back in your lane and keep moving forward. Always.

Forge Your New Path

Prepare for Recalculation by Clarifying Your Goals, Assessing Your Strengths and Educational Credentials, and Better Managing Your Time and Energy

Spectacular achievement is always preceded
by unspectacular preparation.
—ROBERT H. SCHULLER

When you're first learning how to drive a car, you're taught some basics that you should apply every time you get behind the wheel, whether it's drive #1 or drive #1,000,000: put on your seat belt, adjust your mirrors, and make sure your dashboard isn't flashing any warning lights. The same principle applies to career success: there are fundamental steps that must be taken to ensure a successful journey, whether you are seeking your first job or your twenty-first.

We will address a lot in this book that is new or different

because of our current pandemic-affected circumstances, but the foundation-laying practices in this chapter apply no matter what the economy is doing, how high or low the unemployment numbers are, or how much automation is impacting your industry. In fact, the more unpredictable the circumstances, the more important these basics become.

"Begin with the End in Mind"

This strategy is habit number two of another one of my favorite personal development books, *The 7 Habits of Highly Effective People*. Whether you've just finished your degree, been abruptly laid off, or are considering an internal move at your current employer, take some time to fantasize about the career you want. You can define the "end" you're imagining in whatever time frame feels right to you. It might be one year from today, it might be ten years from today, it might be the day you retire, or it might be much later on when you're taking stock of what you've achieved in your lifetime. What's important is settling on a destination to aim for.

Trust your first response to this question: *What does a successful career feel like to you?* Are you at a desk? On a stage? In a hospital? A lab? A courtroom? A classroom? Are you in a big city? Are you in another country? What are you wearing? What's on your to-do list? Are there people around you and, if so, who are they and what are they doing?

Try to make this hypothetical scene of the "successful future you" as detailed as possible. If you're artistic, get a piece of poster board and create a collage of images and words, often called a vision board, that reflect the career you want to build. If you enjoy writing, turn this imaginary scene into a story of an ideal day in your dream career. If you've taken the long view, another option is to write your vision in the form of your future obituary, listing your most notable personal and professional experiences and accomplishments at different stages of your life.

Even a short-term "end" can be incredibly powerful in helping to guide you. A friend of mine worked on the very early days of Barack Obama's first presidential campaign, which was considered quite a long shot back in 2007. Campaign strategists determined that the most important factor in launching Obama to the presidency was for him to win the very first primary contest, the Iowa Democratic presidential caucus, in January 2008. According to my friend, from day one, whenever anyone asked a question about any strategy, action, budget item, or communication, the response was always, "Will it help us win Iowa?"

When you have a clear, unequivocal goal, you have a destination to navigate toward in every moment.

Even if you're undecided about exactly what career you want to have, start with what you *do* know. For example, maybe you don't know which industry you want to work in or exactly what job title you'd like to have, but you do know that you want to work with smart colleagues doing something that helps people. Then, for any decision—which booths to visit at a virtual job fair, what continuing education classes to take, what words to use on your LinkedIn profile—you can ask yourself: "Which events/classes/words will get me closer to my goal of working with smart colleagues doing something that helps people?"

Taking some time to clarify your desired destination—or

at least elements of it—will set you up for success in many of the areas we'll address in future chapters.

Exercise: Enter Your Destination

Use this space to envision your dream career outcome. Whether it's a short-term or long-term goal, whether you have a vague concept or know exactly what you want, put it in writing here. If you need some inspiration, revisit the profiles of people you envied in the previous chapter. You can use this space to write a story, a stream-of-consciousness paragraph, or even just a list of bullet points containing random elements you'd like your next career opportunity to have. The important thing is to clarify your goal and keep this destination in mind over the course of this book and beyond.

_____

_____

_____

_____

_____

Embrace Creative Career Path Shapes

Imagine that you've just entered your dream destination into a GPS. The next step is for a route or path to be calculated in order to arrive there from where you are now.

What image of that path comes to mind for you? If you're

like many people, it's probably a long, flat highway or the pro-verbial career ladder—a series of junior, lesser-paid positions that increase year by year to more senior, higher-paying roles. Please erase this image from your mind if it doesn't serve you! For many people today, the idea of a straight-up, ladder-like career progression is a relic at best and a dangerous red her-ring at worst. Not everyone's career can or will follow this model, and the ladder image ignores the reality of life events, bad economies, and technological evolution. (As just one data point, Pearson found that 73 percent of people globally be-lieve that the notion of working for one employer for your entire career is old-fashioned.)

Instead, know that you can envision your career path to take any shape you choose. Here are some examples to spark your imagination:

Lattice

The "career lattice" concept was conceived by Cathy Benko, Molly Anderson, and Suzanne Vickberg of Deloitte in 2011. In math, a lattice is a three-dimensional structure that ex-tends infinitely in any direction. In work parlance, the lattice metaphor represents a career that is "multidirectional, flex-ible, and expansive." Movement can be up, down, across, or diagonal, and the options are limitless. The lattice can be a particularly helpful career model for people transitioning out of the military into civilian careers. As Michael Abrams, CEO of FourBlock, a nonprofit that provides career-transition sup-port for veterans (and for which I serve as a board member), tells veterans, "Sometimes you have to take two steps back to go a thousand steps forward."

Jungle Gym

Sheryl Sandberg, chief operating officer of Facebook, chose the metaphor of a jungle gym to describe career paths in the modern economy. Like a lattice, a jungle gym shape encourages movement in all directions—up, down, and laterally—and, importantly, jungle gyms are fun and encourage playfulness, stretching, and creativity. There is no "wrong" way to move on a jungle gym.

Portfolio of Investments

Executive coach Nihar Chhaya encourages people to view their careers as a portfolio of investments. In a 2020 article for *Harvard Business Review*, he wrote, "When you identify your self-worth by your job, you are setting yourself up for disappointment anytime you see someone surpassing you. But by defining your career as a diverse portfolio of time and talent investments you make to bring value to the world, you can hedge against the painful moments of feeling behind in some areas, with activities where you build a lead."

Portfolios can include any combination of job functions, industries, roles, and time commitments that don't necessarily have to relate to one another. Similarly, author and blogger Anne Bogel imagines the various aspects of her career like the cars on a Ferris wheel, with the elements always in motion and her focus on each one revolving over time.

Assess Your Strengths and Skills

Now that you have a sense of your internal destination and a wider variety of paths to get you there, let's enhance that plan with some additional research and self-discovery. Your personal vision is incredibly important, but it is limited by the extent of your existing tools and knowledge. I often tell people at the beginning of a job search that it is quite likely they will end up acquiring a skill, or working for a company, or in an industry, they haven't even heard of yet.

An effective way to expand your universe of options is to take an assessment test. Assessment tests can help you cast a wider net by showing you how your natural abilities and interests match with different jobs and careers. There are many to choose from, such as the Myers-Briggs Type Indicator (MBTI) and the Motivational Appraisal of Personal Potential (MAPP), and all will provide value. Check with your university's career center or a local job training center to learn which assessments they might offer at no cost. There are also free or low-cost online assessments, like 16Personalities.com.

My favorite assessment is Cappfinity's Strengths Profile, which takes a strengths-based view of career success. This means that rather than analyzing your weaknesses and finding ways to minimize or overcome them, you instead double down on the skills and attributes that come more easily to you—what you're good at *and* what you enjoy. The assessment then helps you match those strengths with the jobs and careers that utilize them.

Nicky Garcea and Alex Linley, the founders of Cappfinity, a company for which I serve as a brand ambassador, shared with me their insights on what assessments can do for you and how you can make the most of them:

Assessments Provide a Language to Describe Your Strengths and Transferable Skills.

While no test will serve as a be-all-and-end-all guide to exactly what career to choose, assessments offer a helpful framework for self-awareness. Many people are pretty good at describing their weaknesses but lack the language to describe their strengths. Sometimes they don't recognize them at all. Assessments provide a way to identify skills and qualities that don't even occur to you to mention to employers, and build your confidence in how much you have to offer.

For example, Alex's wife has a key strength of "esteem builder," which means that she excels at using her words and actions to help build self-confidence and self-esteem in other people. This strength matches beautifully with her career as a social worker. But before she took the assessment, she had never described herself as having this strength. "I thought everyone was good at that!" she said. Because esteem building came naturally to her, she didn't think the skill was anything special to promote about herself.

Nicky and Alex witness similar aha moments in parents returning to the workforce after taking time out of the paid workforce. An assessment test might identify a strength like "time optimizer" that the parent acquired or strengthened while taking care of their children. The assessment gives the parent professionally applicable language to articulate the benefits of their career break to potential employers.

Assessments Help You Rediscover Your Innate Interests.

Sometimes, assessments will guide you back to activities you loved as a child that might have gotten lost during your formal education but would be great career options. As Nicky points out, many people choose a career based on the subjects in which they excelled at school, such as pursuing accounting because you received good grades in math, not because you actually wanted to be an accountant. Circumstances like a layoff can give people a chance to pause and reflect on the path they've chosen, sometimes for the first time in years.

Nicky and Alex told me that one surprisingly common result of assessment testing is when people who've been laid off from supervisor roles realize that they never actually liked managing people. So often, when you're good at a skill like sales, data analysis, or design, you're promoted to a role where you do less of the work you love and instead manage other people doing that work. An assessment can remind you that just because you've managed people in the past doesn't mean you have to do so in the future.

Assessments Remind You That People Change.

"The generally perceived viewpoint is that personality doesn't change," says Alex. "This is not true. Of course we have predispositions, but there are also situational influences on our strengths." He uses himself as an example. Alex is a former academic and, for a long time, writing was one of his top strengths. That is, until he got to the point where he became really tired of writing. It moved from being a strength that energized him to a learned behavior that drained him. To feel happy and fulfilled in his career, he decided to de-emphasize writing and dive more deeply into other pursuits. While Alex points out that you can't pick and choose your strengths and

Quick Recalculation: "How Do I Find My Purpose?"

The reason I like Cappfinity's Strengths Profile is its focus on assessing not just your innate skills but the work you most enjoy. Building on that foundation, another way to expand your recalculation options is to follow your bliss, also known as "finding your calling," or "living your purpose"—concepts that sound amazing but sometimes can be difficult, and even stress-inducing, to put into practice.

I reached out to Zach Mercurio, a professor and researcher at the Center for Meaning and Purpose at Colorado State University, for guidance on tactical ways to find truly fulfilling work. He understands that people feel a lot of anxiety about finding a singular purpose and worry that something is wrong if they can't do so. The search for a job with purpose or passion can create more disappointment and low self-worth if you don't find "the one" that gives your life significance.

Zach's advice? "Instead of trying to 'find your purpose,'" he says, "it is more important and impactful to focus on 'being purposeful.'"

Zach defines "being purposeful" as concentrating on the contribution you are making, which can take place even when you're unemployed or in a period of transition. Instead of asking, "What should I do with my life?" Zach recommends asking, "What can my life do for others? How can I contribute right now? What problems can I help solve?" There are many ways to answer these questions, so you are more likely to identify multiple ways you might contribute, rather than a single, ultimate calling.

Defining your contribution can be another form of assessment.

Zach recommends keeping a purpose journal, where every day for a week you write down answers to three questions:

1. What was I good at today?
2. What did I love doing today?
3. How did I positively contribute to another person today?

If you write these lists for a week straight, you'll end up with twenty-one answers to analyze. What themes do you observe? What opportunities can you find? The goal is to create a sense of self-belief that you have something to contribute right now, no matter what your situation. "We are so good at telling ourselves what we should become," says Zach. "Instead, start to notice what's good about who you are and your unique abilities right now."

You can apply the same approach if you find yourself in a job that is less than ideal. Zach's research has found that purposeful work is accessible in truly any position. For one research project, he studied janitors working at a university. He found that some of the janitors were miserable and others were quite happy. They all told Zach they had taken the job because they needed a paycheck, but the happier janitors found ways to uncover a sense of purpose in their work. One janitor would tell herself every day, "I'm cleaning the bathroom so that these kids don't get sick." That's a purposeful attitude.

When you frame your job in terms of your positive impact on other people or the world, you are bound to feel more positive about it. As Zach says, "The experience of meaningful work has less to do with what you're actually doing and more to do with your approach to what you're doing."

weaknesses, an assessment can make you aware of how your interests and skills evolve and change over time.

In quite a positive side note, Cappfinity data found some encouraging changes in the strengths of the overall global population throughout the COVID-19 pandemic. They measured a global rise in the strengths of gratitude and service as a result of our shared struggle with the virus.

Nicky and Alex have generously offered a free version of Cappfinity's Strengths Profile to readers of *Recalculating*. You can take the assessment at lindseypollak.com/SP.

Evaluate Your Education

While I encourage you to think as creatively as possible about your career path and ways to make a living in a meaningful way, your level and type of education are likely to play a role in determining your options. Let's explore how the education market is changing and how college, graduate school, and non-degree programs may fit into your path, now and in the future.

We'll begin with the fact that more Americans have a college degree today than ever before: 33.4 percent of the U.S. population aged twenty-five and above have graduated from college, compared to just 4.6 percent in 1940. However, this does not mean that a third of Americans spend four post–high school years studying and frolicking on the campus of a leafy college quad to obtain a degree that will set them up for lasting success. That idealized image is far from the norm today, similar to the career-ladder fantasy we discussed earlier in this chapter. While overall attainment of college degrees is on the rise, consider some facts:

- Six out of ten people who enroll in college don't receive a degree within six years.
- 38 percent of undergraduate students in the United States are over the age of twenty-five.
- 70 percent of full-time college students are working in addition to studying, with 26 percent of those working full-time.

The outlook for the post-pandemic future is that even more change will occur to the traditional notion that a post–high school four-year college degree is sufficient education for a lifelong career. "Your children can expect to change jobs *and professions* multiple times in their lifetimes," wrote *New York Times* opinion columnist Thomas Friedman in October 2020, "which means their career path will no longer follow a simple 'learn-to-work trajectory,' as Heather E. McGowan, co-author of *The Adaptation Advantage*, likes to say, but rather a path of 'work-learn-work-learn-work-learn.'"

Manny Contomanolis, senior associate vice president of employer engagement and career design at Northeastern University, told me that Northeastern now uses the term "learners" rather than "students" because education can now be a lifelong endeavor achieved in a multitude of time frames and environments. All of this is to say that if you are, or are considering becoming, a "non-traditional" college student, you are far from alone.

It's hard even to begin a conversation about education in

the United States without addressing cost, whether you are currently paying tuition or student loans or you are considering taking on this cost. As has been well reported, higher education costs today are astronomical and burden too many young people in particular with significant debt before their careers have launched.

Even before the pandemic-induced recession, a backlash had begun against the high cost of higher ed and the perception that a person cannot have a successful career without a college degree. Younger Americans, not surprisingly, are leading the rebellion. One study in 2020 found that just 23 percent of Generation Z high school students believe a four-year college is the path to a good job. One reason why that might be the case is that many employers in the popular tech sector—Google, Apple, and IBM among them—are no longer requiring a bachelor's degree for many jobs that used to require one.

Still, for the vast majority of professional careers today, a bachelor's degree remains a must-have credential. It is also the surest path to financial security. College graduates now earn 56 percent more than high school graduates, which is the widest earnings gap between the two groups on record.

So how should higher education factor into your recalculation path? We'll consider the topic from a variety of perspectives. There is certainly no reason to regret a four-year degree if you have one, but you might need to acknowledge it's not enough to achieve your recalculation goals. The good news is you have more options for learning than ever before. Like so many other realms, the industry of higher education is in the midst of significant disruption that was accelerated by the pandemic. In addition to traditional in-person colleges, universities, and community colleges, people of all ages can access college-level coursework through online methods including massive open online courses (MOOCs), certificate programs, executive education institutes, and private organizations like General Assembly. The overnight move to remote learning that took place with the onset of the pandemic will surely lead to even more disruption in the sector.

Whether you are in the market for a new job or seeking to pivot in your current organization, I encourage you to embrace a "lifelong learning" mentality. It is critical to stay agile and continue to learn and develop your skills as technology continues to advance and change our economy. "Learning is the new pension," says Heather E. McGowan. "It's how you create your future value every day." Many people came to this conclusion in the pivotal summer of 2020: after the first few months of the pandemic had wreaked havoc on the job market, a national poll by the Strada Education Network found that 37 percent of eighteen- to twenty-four-year-olds and 23 percent of twenty-five- to sixty-four-year-olds planned to enroll in an education or training program in the next six months.

With the help of some experts, I'll answer some common questions about education for recalculators at various career stages:

Q: As a current student, does my college major matter? If it does, what major should I choose?

A: The answer to this question depends on the industry you want to pursue. There are some professions, like accounting and architecture, where you might need that particular major to land a job. But in the vast majority of career and job search situations today, your major really doesn't matter. The reason is that your education is about much more than the subject matter of your classes. Here are some of the things that employers tell me are more important than a college major when they are assessing job candidates:

- **Experience:** Virtually every employer mentions experience as the most desirable résumé item a job candidate can offer. It doesn't matter whether that experience comes from internships, extracurricular activities, volunteer work, part-time jobs, gig work, independent study, or working a minimum-wage job.
- **Job-specific skills:** Hand in hand with experience come tangible skills. If you want a programming job, for instance, your performance on a sample coding project will matter much more than the words on your diploma. The same goes for artistic talent, sales ability, foreign language fluency, or any other measurable skill that is required for success in a particular position. Such skills can be inborn or learned outside the classroom as much as in it.
- **"Soft skills":** These include capabilities like written communication, presentation skills, critical thinking, relationship building, problem solving, and teamwork. You can build these skills in any college major. We'll explore these more in Chapter Three on crafting your career story.
- **Connections:** The more you build a diverse network in

college or any other learning environment—by forming relationships with classmates, professors, advisers, career services professionals, internship colleagues, alumni, and others—the more career opportunities you'll have no matter what your major or course of study.

Where a college major *does* matter is what it says about where you decided to focus your time and energy during your undergraduate experience. It is an indication of your interests and the knowledge base you wanted to build. This is what employers are often interested in rather than the actual content of your classes. The reasons why you chose a particular major often say more about you than the major itself. What is it about the major that is interesting to you? How did it fit into your larger career and life goals?

If you are a current student and haven't yet chosen a major, my best advice is to pick something simply for your intellectual interest in the content. In many cases, following your genuine interest with no career-related goal can actually have a professional benefit. It might just take a while for this to happen. Here's a great example: Karen Ivy, director of career services at Texas A&M University–San Antonio, once had a student named Will who, when reflecting on his chances of landing his desired job in wealth management, told her that he regretted taking classes on Russian history and Russian literature just because he enjoyed them. "What was the point?" he lamented. Years later, Will reconnected with Karen to share the news that he had just landed a big client from Russia. Will told Karen that his genuine interest in, and knowledge of, Russian culture is what won him the business. You just never know how something you studied for no reason other than pure interest might become the link to a positive career experience.

Here's another perspective on college majors to consider from Julie Lythcott-Haims, former dean of freshmen at Stanford University, who shared some unique advice on an episode of the podcast *The Anxious Achiever*. She says that as an adviser, she used to focus less on a student's major and, instead, "root for the minor." Here's why:

"Students would come to my office hours and I would be getting to know them. They would say, 'Well, I'm majoring in econ, or pre-med, or engineering, whatever. They'd smile in that sheepish way or shrug their shoulders. And that was their way of saying *because I have to*. And then they would say, 'I'm minoring in film studies.' Or, 'I'm minoring in English literature.' Or, 'I'm minoring in photography.' And I would beam when I heard about the minor. I always thought it was my job to elevate the status of the minor, because the minor was often where the student's actual sense of self lay."

Remember getting rid of the "shoulds" in Chapter One? Take some time to think about whether you are choosing, or have chosen, a college major not because it is what you want to study or pursue as a profession but because it is what a parent, friend, professor, or news article deemed a good idea. Maybe your more intuitive choice of a minor, an elective, or an extracurricular activity will be a more valuable predictor of the path you should pursue for career fulfillment.

This was certainly true for me when I recalculated my own major after my sophomore year of college. My dad is a retired

English teacher and I was a strong English student in high school, so I had just assumed I would major in English in college. I didn't love my English classes my first year—think *The Iliad, The Odyssey, The Aeneid*—and then sophomore year when I started taking the more intense courses—think major English poets, Shakespeare, Shakespeare, and, wow, there was so much more Shakespeare—I began to actively dislike my classes.

Second semester sophomore year, I somewhat randomly signed up for a class called Formation of Modern American Culture, which was an American Studies course that covered the intersection of literature, history, politics, music, art, and popular culture in the 1800s and 1900s. I absolutely loved it, devouring every lecture and assigned book and even diving into the "supplemental reading" (which I had never, ever done before). At first, it didn't even occur to me that I could change my major to take more classes like this one. The problem was that American Studies was considered an easy or "gut" major, certainly not as prestigious as English. Everyone referred to it as "Am Stud," if that gives you a sense of its reputation.

I really struggled with the decision of whether or not to change my major. I thought I "should" major in English. Ultimately, it was a conversation with a senior American Studies major that finally inspired me to make the change. We met for coffee, and when she told me about the classes she was taking at her level, I felt an excitement and interest level I had never felt about my English classes. The aha moment reminded me of *When Harry Met Sally*: "I'll have what she's having."

I officially changed my major soon after.

Looking back, I think this decision was the first I made

in my adult life that involved turning away from the "presti-gious" or "expected" option in favor of trusting my own in-stincts. (And I acknowledge my deep privilege in having such options available to me in the first place.) When in doubt, trust your gut.

Q: What should I do if I have a low GPA?

A: "How to Get a Job When You Have a Low GPA" has been the number one most read post on my blog (lindseypollak.com/blog) for more than ten years. If you're a student or recent grad concerned about your GPA, I want to set your mind at ease. You absolutely can find a job despite a less-than-stellar academic record, even in a time of high unemployment. But you do have to be realistic and strategic.

Here's the scoop: some elite employers do have policies re-quiring a certain GPA (usually a 3.0 or higher), and there is generally no way around that rule. So, if you have a low GPA, do your research on any potential job opportunities to learn if your GPA takes you out of the running. Employers with clear cutoffs will state their rules in their job postings. If you don't meet their criteria, then move on to the zillions of other em-ployers that don't have a cutoff. Remember recalculator rule #3: control what you can and let the rest go.

The good news is that GPA cutoffs are less common than they used to be. Even Google, which used to be famous for ask-ing candidates of all experience levels for their college GPAs and SAT scores, finally stopped that practice, aside from some entry-level roles.

That being said, you do need to have a plan for how you will address your low GPA if an employer asks about it. First, understand why an employer would ask about your grades, which will be most common if you're a recent college graduate. When you don't have much work experience, your educational results provide an employer with some information about how you operate. Your grades are often perceived as a reflection of your diligence and your work ethic.

I've heard of job offers being rescinded because of a potential employee's low grades during their second semester of senior year. Employers don't want to hire people who only work hard "when it matters." So, if your grades aren't that great, it can be helpful if you've at least made some improvement in your GPA over the years and can point out that progress to employers.

Another way to handle a low GPA is to demonstrate that you've earned better grades in your major, your minor, or in classes related to the job you're applying for. For this strategy, you can list your "major GPA" or "grades from relevant classes" on your résumé if you'd like, or just mention this in a cover letter or during a job interview.

If your grades are low across the board, I recommend that you prepare a short, honest explanation for when a potential employer questions your low GPA. For instance, "I struggled academically in college, but I was fully committed to my extracurriculars and would love to tell you about some of those accomplishments." Or, "I was dealing with some family health issues sophomore year and that affected my grades. It was very challenging, but I learned a lot from the experience." Or, "I paid my way through college, so unfortunately my grades suffered sometimes because of my work schedule. I wish my grades had been better, but I'm proud that I was able to support myself all the way through." Of course, you'll want to think about the level of information you're comfortable providing; just make sure you have an answer ready if a potential employer asks. It's also a good idea to have a few letters of recommendation confirming your intelligence and work ethic, perhaps from an employer or extracurricular adviser.

Here's the most important thing to keep in mind about your GPA: it matters less and less as you advance in your career and have more experience under your belt. This means that if you want your GPA to matter less, you need to make your experience matter more. On several occasions I've been impressed by eager, ambitious, engaging young people who later told me they had a low GPA in college. Once I liked and trusted them, their GPAs didn't matter at all.

And by the way, do not—I repeat, *do not*—bring up your low GPA if an employer doesn't ask about it. Many potential employers will not bring up your grades, and if that's the case, you are under no obligation to reveal them. You are also not required to list your GPA on your résumé or LinkedIn profile. In fact, I'd suggest only listing a GPA of 3.0 or higher, only for your first few years out of college, and only if you want to.

Q: Should I go to grad school or take a gap year?

A: Let's take these questions separately, starting with graduate school. Grad school attendance typically rises during recessionary times, as people think it's a good way to "ride out" a bad economy. But is that a good decision?

"I don't think there is a 'right' answer," says Michelle Horton, executive director, career education and coaching, market readiness and employment at Wake Forest University School of Business. "Everything comes back to personal choice and decision. The first question to ask yourself is: Can you afford to go to grad school?"

I completely agree that cost is a major factor for deciding when to obtain a graduate degree, where to obtain it, and which degree to pursue. Personal finance expert Ramit Sethi advises potential applicants to factor in the opportunity cost of grad school along with the tuition and living expenses. "Not only are you going to be paying for this program, you're also going to be losing money by not being employed," he says. Ramit also cautions that some degrees are worth more in future earning potential than others. He says that it's not worth seeking a graduate degree if the tuition and expenses cost more than the first year's starting salary in the field.

If you were already planning to go to graduate school before the pandemic or if you need an advanced degree to pursue your chosen field (such as becoming a professor or doctor), then by all means move forward. I would also say that you should definitely go to graduate school if your current employer will pay for all or part of it or you receive a scholarship or fellowship to do so and it's an experience that you want. (This is how and why I got my master's degree. I am forever grateful for the Rotary Ambassadorial Scholarship.)

When it comes to taking a gap year during college or right after graduating, the same considerations apply. Can you afford it? What is the opportunity cost? Was this something you had already considered or are you just taking time off because you are afraid it will be too hard to get a job?

During the summer and fall of 2020, many students faced these big decisions. I asked Michelle Horton how she was advising students considering a gap year in a time of such upheaval. She said it's the same advice she gives students in any period of uncertainty or indecision: you must have a plan for how you will spend your gap year productively. "What I *do not* recommend," she told me emphatically, "is doing nothing. Always show activity, engagement, and a willingness to learn."

Michelle told me the story of one student who decided he wanted to go into digital marketing but did not have some of the specialized skills that were required in this space. During school breaks and weekends, he took a few courses on digital marketing using the LinkedIn Learning platform, listed that on his résumé, and was able to land an entry-level job in the field when he graduated.

Q: How do I know if an additional degree or certificate is necessary to my recalculation?

A: At any stage of your career, the answer to this question is to do your research. Scan online job postings for positions that interest you and make note of the qualifications. If you're a career changer, ask anyone in your professional network who works in the industry you're seeking to join if additional education will help your employment prospects (check out Chapter Four for more advice on conducting informational interviews). Ask professionals from your university career center about educational requirements related to various career options as well, and don't hesitate to inquire whether a certain number of years in the workplace might make your lack of degree a nonissue. You're bound to notice a recurring theme if certain proficiencies are expected—or can be skirted.

You don't want to invest time and money in an unnecessary degree, but you also don't want to start applying for jobs you have no chance of obtaining without certain credentials. If you're on the fence and you can find an affordable, short-term option, then go for it. It's never a mistake to learn something new, and enrolling in more education demonstrates to employers that you are committed to lifelong learning by either building a new proficiency or refreshing existing competencies. Even a single online class can make a difference.

Q: If I determine that I need more training (but not an additional degree), what are the educational options I should consider?

A: Whether you need to brush up on hard skills like mastering a new technology or soft skills like becoming a better manager, there are plenty of options—and many are available remotely or on flexible schedules. Check out platforms like LinkedIn Learning, edX, FutureLearn, and Coursera. Community colleges, along with four-year colleges and universities, also offer a wide range of fields of study through certificate programs and executive education. (Visit Chapter Three for more specific guidance on reskilling and upskilling.) I am also a big fan of one-off webinars to tap into education-on-demand for any topic imaginable. Industry associations are particularly good resources for finding these shorter, online, low- or no-cost educational opportunities.

Q: What are some options for more experiential training?

A: Not all learning happens in a classroom or on a computer screen, and some people learn best by doing. One great place to start is volunteering for a nonprofit to gain additional skills and experience; for example, learning more about social media by offering to manage the Instagram account for your local food bank or polishing your sales skills by approaching donors on behalf of your child's school.

You also could try an apprenticeship or job shadowing, where you spend time with an established professional to find out more about what they do—and to determine whether that job would be a good fit for you. This type of experience can be formal or informal; check with your alma mater's career services office to find out if they maintain lists of fellow alumni willing to help job seekers and career changers, or simply ask a friend if you can tag along for a day.

And of course finding an internship is another great option—and it doesn't even have to be a full-time or in-person gig. College career center job boards, entry-level-focused sites like Handshake, and the micro-internship organization Parker Dewey are all good options to find an internship. If you're a mid-career recalculator who has been out of the work world for a while, you might seek out a "returnship," which is an internship-like program specifically designed for those in your shoes. Many large companies offer such programs and promote them in the career section of their websites, on public job boards, or on the dedicated returnship website iRelaunch.com.

Finally, no matter what your age or experience level, you can get a little gutsy and invent your own internship or returnship experience. I came across an incredibly inspiring story of this type of invention from Morgan Peterson, a rising senior at North Carolina State University, who is majoring in fashion and textile management with a concentration in brand management and marketing.

Morgan had always assumed she'd spend the summer of 2020 between her junior and senior years holding down a corporate internship at a big fashion label in New York City or London. But COVID-19 had other plans for her. Undaunted, she decided to create her own internships, combing Instagram for companies and brands she'd most want to learn from and contribute to.

She sent "cold" emails to eighty people, which included this paragraph:

I am reaching out to express interest in assisting your company during the pandemic. I understand that now is a stressful time for businesses as circumstances are changing daily. I believe that I can assist your company in marketing, social media management, and customer relationship management. I am very flexible and adaptable which should be an asset in this new normal in which we are living.

The result? Morgan landed four PR and social media internships in fashion brands in the United States and Rome. This young woman is a fantastic role model for a recalculator of any age.

Quick Recalculation: Should I Start My Own Business?

Still not sure how to begin your recalculation path? Let's concentrate for a moment on the option of self-employment, including freelancing, gig work, consulting, and starting a small business. The great news is that entrepreneurship is truly "equal opportunity": in the words of one of my early mentors, "All it takes to start a business is a pulse."

Rest assured you don't have to be a young techie coding in your garage to be an entrepreneur; if you're an older recalculator, you might be pleasantly surprised to learn that the average age of someone starting a new venture is forty-five, according to the *Harvard Business Review*. You also don't have to seek venture capital investment or take out a large business loan to start a company, although that is, of course, a potential route to take. Plenty of people launch businesses just by setting up a website or online profile.

The approach I recommend for people considering self-employment is to start by dipping a toe in the entrepreneurial waters. As you know, I started my business somewhat accidentally while I was also seeking full-time jobs, so my best advice to most recalculators who aren't 100 percent sure they want to be self-employed is to pursue jobs and entrepreneurship at the same time and notice what sticks.

In the modern working world, there can be much more fluid movement in and out of employment, self-employment, contract work, and beyond. Inside some companies, workers of different statuses all work on projects and teams together.

For more guidance on launching into self-employment, my favorite resources are SCORE.org, allBusiness.com, and the Freelancers Union.

Make More Time for Recalculating

Whenever I'm on a time-sensitive car ride, like running late to a meeting, I hold my breath while the GPS app calculates the estimated driving time to my destination. The projection is subject to change, but there is always a minimum required time. The same goes for job hunting and career growth activities.

A common mistake I observe among recalculators is the belief that you can simply "fit in" your career-development activities around the rest of your life. Many people don't achieve their career goals because they don't make time to pursue them. I urge you to schedule your recalculation-related activities into your calendar and add them to your daily to-do lists.

If you're recalculating while in school or working part-time or full-time, you'll need to block off hours in your already busy schedule to focus on forging a new path. If you are job hunting or launching into self-employment full time, you'll need to create structure to make the most of your days. Laura Vanderkam, host of the remote work podcast *The New Corner Office* and the author of multiple books including *Off the Clock: Feel Less Busy While Getting More Done*, is an expert on all aspects of time management and productivity, so I turned to her for some suggestions for recalculators in both camps. Here are her top three tips:

Wake Up Earlier.

One of the best ways to fit anything into an already full schedule is to get up earlier. If you normally spend late-night hours surfing the Web or watching TV, try to readjust your sleep schedule. By going to bed earlier, you can turn unproductive evening hours into productive morning hours. Give yourself a list of job-hunting or career development activities you can do in the morning, and then carve out a little bit of time during the workday for things that need to happen during work hours. You'll make progress!

Set Boundaries Around Your Career-Related Activities.

Activities like job hunting, launching a freelance career, and pursuing further education can expand to fill all available space in your day. Since these activities can also be stressful, you're best off figuring out a way that you can officially be "done" each day. Give yourself a set list of tasks or a set number of hours to spend focusing on your pursuit. When that's over, you're done. You can relax and do other things without constantly feeling like you should be perusing job listings, perfecting assignments, or sending networking emails.

Manage Your Energy by Proactively Planning Breaks Throughout the Day.

A good rule of thumb is to plan for three breaks during a normal workday (mid-morning, lunch, and mid-afternoon). Aim to make one break physical and one social (this can include being virtually social, such as calling or FaceTiming a friend or relative). If you are working, studying, or job hunting from home, you also might find it useful to have a shutdown ritual. A notable problem with working from home is that there's no obvious endpoint, which a commute would give you if you were working elsewhere. So come up with your own end-of-day transition. Maybe you can jot down your to-do list for the next day, write in a journal, or sign off with a particular friend or fellow job seeker.

Prepare for Bumps in the Road

No matter how well you manage your time, the truth is that recalculation can often take longer than you want it to, even if you do all the right things. It doesn't mean you're slacking or messing up; it just means you're taking longer than expected to land a job, change careers, receive a promotion, or make whatever transition you desire.

Instead of bemoaning the ups and downs of this journey, I suggest deliberately planning for the roller-coaster nature of it all. One of the tricks to handling adversity, boredom, or disappointment is to not be surprised by it. No matter how perfectly laid out your career plans are, life happens. (And, for what it's worth to millennial and Gen Z readers, I think it's important to acknowledge that today's younger generations have faced a disproportionate number of societal and economic "happenings" in the early years of your careers.)

Each person faces their own unique set of internal and external challenges, and some of us have more privilege than others due to the color of our skin, our sexual orientation, our social class, our physical abilities, our family history, and more. But everyone faces difficulties. So let's conclude this chapter on preparation by preparing two strategies to handle those inevitable bumps in the road.

Lower Your Expectations

A colleague in one of my first jobs shared with me that the secret to her success was low expectations. "I basically expect about 30 percent of my day to suck," she told me matter-of-factly. "So, if I'm only annoyed or frustrated for, like, 25 percent of the day, then it feels like a win!"

At first I thought this sounded depressing, but then I realized it was brilliant. If you expect perfection, you will almost always be disappointed. But if you expect imperfection, you will be less discouraged when it happens—and thrilled when those expectations are exceeded! Lowering your expectations doesn't mean giving up on your dreams; it means giving up on the belief that total perfection is possible. It means setting realistic goals. For example, if you send ten emails to alumni of your university and expect to hear back from all ten, you'll be disappointed if only nine respond. If you send those same ten emails and hope for one or two to reply, you'll be ecstatic if you hear back from four.

Build Momentum

From March to April 2020, my speaking business came to a standstill in a matter of weeks. All my booked speeches canceled and many clients asked for refunds on deposits they had paid. No one knew how long it would be before live events took place again, and no one felt ready to book online programs yet. We also knew very little about the virus and I was living in New York City, the epicenter of the pandemic at the time. I was grateful to have my health, the health of my family, and my savings account. But, like so many other entrepreneurs, I was also very frightened about the future of my business.

Despite that fear and the abrupt halt of business during that time, I knew it was important to do little things to try to advance my professional goals—remember recalculator rule #2: prioritize action. I sent check-in emails to find out how clients were faring. I commented on the LinkedIn posts of fellow authors, consultants, and speakers. I wrote how-to articles with tips on working from home. (Note that I did these things while also eating my body weight in peanut M&M's every day, so I'm not saying my process was perfect in any way.) The goal was to keep pushing myself forward, even the tiniest bit, despite the incredibly challenging circumstances. Slowly but surely, speaking inquiries began to flow back in.

Starting with merely the smallest action has a dual benefit: First, you get the benefit of whatever action you take. Second, you get the benefit of momentum. It is much easier to take a second, third, or fourth action once you've taken the first one. So decide in advance that whenever you face a bump in the road, you will take a tiny action toward your goal. Momentum and consistency are secret superpowers. In the words of Ovid, "Dripping water hollows out stone, not through force but through persistence."

Exercise: Momentum Builders

Here is a list of small actions that can help you build momentum in a variety of ways, depending on your unique goals. Come back to this list whenever you feel overwhelmed by challenges or frustrations on your recalculation path and you need to kickstart your momentum again. And there is no reason to stop these career-building actions even after you achieve your current goal. Imagine where you can end up if you just take one little action every single day, now and forever?

❑ Improve one sentence on your LinkedIn profile.
❑ Add one new key word to your résumé that you've noticed in job descriptions that appeal to you.
❑ Attend one webinar on job hunting, entrepreneurship, or any professional topic that appeals to you.
❑ Declutter one item from your workspace or background to create a more professional environment for video calls and virtual meetings and interviews.
❑ Listen to one podcast episode related to your current industry or an industry you'd like to work in.
❑ Research a list of most common job interview questions and draft your answer to one of them.

❑ Expand your distance search criteria on a jobs site by a five-mile radius.

❑ Follow one potential employer or client on Twitter or LinkedIn.

❑ Google and read one article or watch one TikTok on how to perform well in a job interview or salary negotiation.

❑ Text three friends to ask how they landed their most recent job or promotion or had a success in their business.

❑ Connect with three former classmates or colleagues on LinkedIn.

❑ Volunteer one hour of time, virtually or in person, to a cause that is important to you.

❑ Request to join one LinkedIn or Facebook group in your industry or desired industry.

❑ Sign up for one new industry- or career-related e-newsletter.

❑ Reach out to one person you know professionally and ask if you can do anything to support them.

❑ Respond to a blog or LinkedIn post by a leader in an organization or industry that interests you.

❑ Send a thank-you note or email to someone who has supported you.

Your Career Story

**Clarify Your Personal Brand and "Soft"
and "Hard" Skills, and Communicate Your
Story Through Your Résumé, Cover Letters,
LinkedIn, and Other Social Media**

Every day we choose who we are by
how we define ourselves.
—ANGELINA JOLIE

Carla Harris is one of my career role models. She is a managing partner at Morgan Stanley, as well as a powerful keynote speaker and business advice author, and she once had these wise words to say: "All major decisions about your career will be made when you are not in the room."

While that's quite a sweeping statement, it's also one I've found to be 100 percent true. If you interview for a job, it's no secret that you'll be evaluated the moment you walk out the door or log off of the Zoom meeting. If you're up for a promotion or raise, you will not be present when your bosses decide your fate. If you pitch an entrepreneurial venture to investors, you may nail it, but you will not be included in the conversation when they discuss whether or not to fund you. Even if you're the CEO of a large corporation, your board will discuss your performance in private.

It doesn't seem fair that you can work really hard and then not be in the "room where it happens" (for the *Hamilton* fans in the crowd) when your own future is at stake. But while you can't change the outcomes of discussions about you, you can influence them by creating—or re-creating—an overall narrative about your career, including a strong personal brand.

My preferred definition of "personal branding" is "the effort to communicate and present your value to the world." I sometimes hear people say they don't like the word "brand" being applied to them because it feels too commercial or manufactured; if this is the case for you, just replace it with the word "reputation" instead.

Building and maintaining your personal brand is critically important for any recalculator, and it's equally important when you are feeling perfectly comfortable and secure in your current situation. It's important whether you're in college or grad school or you've been working for fifteen or forty-five years. It's important whether you are an employee or an entrepreneur.

The reason I say this so emphatically is the changes that have arisen in the workplace over the past twenty years. Up until the twenty-first century, it was often enough for ambitious professionals to affiliate themselves with a particular organization or industry to be successful: "I work for Ford" or "I'm a nurse." At a time when most people worked at the same organization for many decades, if not their entire careers, having an employer or professional affiliation sufficed to keep a person's career moving forward. If you were an employee, you didn't really need a more specific personal brand because you likely didn't want to leave your employer. Meanwhile, they also wanted you to stay and assimilate with the organization's brand. Depending on your position and performance, they usually had a plan for your advancement and even provided you with a pension to help fund your retirement. This meant that individuals had much less control over their exact career movements but much more security.

Those days are gone forever.

Today, employer-employee relationships are more short-term. Pensions are rare. Career changes are common. Technological advancement can alter entire industries and economies. If you don't act as the CEO of your own brand and career, nobody else will.

As CEO, you are responsible for creating and marketing your brand to others. Just as Coca-Cola works hard to be perceived as refreshing, Volvo works hard to be perceived as safe, and Walt Disney World works hard to be perceived as the happiest place on earth, you need to project an image of yourself to steer the actual words people will use to describe you and to make decisions about engaging with you.

In the previous two chapters, we talked a lot about differentiating between what you can and can't control as a recalculator. Building and communicating your personal brand is firmly in your control.

Your Career Story, Part I: Defining Your Personal Brand

In the first half of this chapter, we'll work to define your personal brand, which includes three elements that we'll review, step by step:

1. Your personal attributes and nontechnical skills ("soft skills"), like being a good leader, a team player, or an effective communicator
2. Your technical proficiencies ("hard skills"), such as an understanding of WordPress, Excel, or Photoshop, which includes your formal educational credentials
3. Your professional experiences

In the second half of the chapter, we'll explore how you communicate your brand and story to the world through "collateral materials," including your résumé, cover letters, and social media profiles.

Step One: Articulate Your Personal Attributes, a.k.a. "Soft Skills"

In business management writer Tom Peters's seminal 1997 *Fast Company* cover story, "The Brand Called You," he explains that "everything you do—and everything you choose not to do—communicates the value and character of the brand. Everything from the way you handle phone conversations to the email messages you send to the way you conduct business in a meeting is part of the larger message you're sending about your brand."

In other words, every action you take in your career provides others with information about who you are. What comes to people's minds when they think about you as a professional? What examples stand out from their various experiences with you? Professional contacts might draw on these memories later on, perhaps when writing you a letter of recommendation, serving as a reference, connecting you to a colleague for a professional opportunity, hiring or promoting you, or deciding whether to lay you off.

I don't say this to scare you into worrying that every single interaction you have with every single person must be absolutely perfect, because it may make or break your reputation and career. I actually say this for the exact opposite reason: to empower you. I say it to remind you that you have infinite opportunities every single day to build and refine your reputation, regardless of where you are in your career journey.

I urge you to be deliberate and proactive. Don't just hope to be noticed for the qualities you want to be known for; take action to make sure you are demonstrating these qualities as often as possible. Every time you send a well-written email, you build a positive impression. Every time you log in on time for a video networking call, you make a positive impression. Every time you leave an eloquent comment on a potential client or employer's LinkedIn post, you make a positive impression. Every time you help a former colleague by sharing a piece of job search advice, you make a positive impression. Every action, no matter how seemingly small, is like adding capital to your reputation bank account. As business author Keith Ferrazzi writes in his classic networking book, *Never Eat Alone*, "Little choices make big impressions."

Exercise: Define Your Key Personal Brand Attributes

Here is a simple exercise to help you attain the personal brand you want to have by defining the "soft skills," or personal qualities, you want to be known for. The exercise consists of three questions:

What Attributes Are You Currently Known For?

If you were to poll everyone who knows you today as a student or professional, what three personal attributes or qualities would most commonly be used to describe you? Remember that the question is not how you would describe yourself (which differentiates this exercise from a self-assessment test) but how you think others would describe you.

Examples: analytical, compassionate, creative, decisive, assertive, diligent, good communicator, ethical, team player, leader, attentive to detail, problem solver, friendly, humble, trustworthy, inquisitive, empathetic

1. _____
   _____
2. _____
   _____
3. _____
   _____

What Personal Attributes Do You Aspire to Be Known For?

In the ideal future you wrote out at the beginning of Chapter Two, where you've achieved a career that makes you happy and fulfilled, what three words would your professional contacts use to describe you? In other words, what is your aspirational personal brand? Perhaps there is some crossover in your words from the previous list, or maybe you have a vision of being perceived with completely different words. There are no wrong answers.

1. _____
   _____
2. _____
   _____
3. _____
   _____

What Regular Habits Will You Build to Advance from Your First List to Your Second One?

Here's where the rubber meets the road. What are you going to do to advance yourself from the reputation you have now to the reputation you want to have? You can't build or enhance your personal brand by thinking really hard. You have to build *habits* that demonstrate these qualities to others on a regular basis.

Take some time here to list three to five regular habits you can build to demonstrate the qualities you aspire to be part of your personal brand. Remember the growth mindset exercise in Chapter One: you can get better at anything if you put in the time and effort.

Examples:
Aspirational brand: leader, detail oriented, excellent public speaker

- To build my brand as a leader, I will listen to one podcast episode a week on a leadership topic.

- To demonstrate that I am detail oriented, I will make it a habit to review and spell-check every email before I press "send."
- To demonstrate that I am serious about improving my skills as a presenter and public speaker, I will attend a monthly Toastmasters International meeting in person or online.

1. _____
   _____
2. _____
   _____
3. _____
   _____
4. _____
   _____
5. _____
   _____

My recommendation is to think of your aspirational image from the above exercise as your personal North Star. No matter what happens during your job search or career development—a layoff, a rejection letter, a less-than-stellar performance review, a challenging boss, an economic downturn, another global crisis—remain focused on handling the situation using the attributes you'd like to strengthen and represent. It might help to write your desired traits on a sticky note that you post in a prominent location. Every morning, look at those words and ask yourself, *How can I find opportunities to express or enhance these attributes today?* Then, try using those ideas to help shape the things you say, think, and do.

When you put those ideal qualities at the center of your thinking and allow them to influence your decision making, you start to slowly alter your normal trajectory. Perhaps the leadership podcast suggests a time management strategy that you implement, which clears up an hour of your day to spend working on a blog post you hadn't previously had time to write. Perhaps your new spell-checking habit catches a supervisor's spelling mistake before she sends out an email publicly and she compliments you for your attention to detail. Perhaps a fellow Toastmasters member watches you give a speech about your career interests and recommends you for a job at his company. Action yields results.

Step Two: Assess Your Technical Skills, a.k.a. Your "Hard Skills"

Now that you've identified the "soft skills" you'd like to be known for, let's address the "hard skills" you'll need to make your desired recalculation. We're talking about "upskilling" and "reskilling," which have emerged as two of the trendiest words in the labor market over the past decade, and became even more ubiquitous when unemployment skyrocketed in the wake of the pandemic.

Upskilling involves learning new and relevant competencies to do your existing job without changing career paths (e.g., a software developer learning a new programming language). Reskilling refers to learning the skills of an entirely new occupation, usually because your current one is becoming obsolete (e.g., a receptionist who retrains to learn Web design). Every recalculator—no, every professional today—needs to consider upskilling or reskilling on a regular basis. I can't think of a single industry or profession today where it's possible for you to determine, "Well, I've learned enough. I'm good."

The driving force behind the need for continuous skill improvement is, of course, the rapid advancement of technology. We are currently in the midst of what economists have identified as the Fourth Industrial Revolution, defined in 2015 by Klaus Schwab, executive chairman of the World Economic Forum, as the era in which technology becomes embedded within societies and even our human bodies. This includes the pervasiveness of technologies like mobile devices, 3D printing, smart sensors, wearable technology, and augmented reality.

The job market reflects these technological advancements. According to a 2018 independent task force report from the Council on Foreign Relations, nearly two-thirds of the thirteen million new jobs created in the United States since 2010 have required higher levels of digital skills. "All workers today need to have a vocabulary around technology," Tom Ogletree, vice president of social impact and external affairs for General Assembly, told me. "Because all companies are now tech companies to one degree or another."

This rapidly growing need for digital savvy among American professionals accelerated at warp speed when the pandemic hit. As Ramona Schindelheim of WorkingNation reported in 2020, "Many [business leaders] believed they had more time to train their current employees to meet the evolving demands of their industries. . . . What was expected to happen over the next 10 to 20 years has, instead, happened in just a few months. The pandemic has upended that timeframe."

What happened is that workers across virtually all industries had to be proficient—or get proficient fast—in digital tools like Zoom, Microsoft Teams, Slack, SharePoint, Google Docs, and more. This transition forced a large number of people to become on-the-job recalculators, and some workers found themselves behind the technology curve.

If you haven't done so already, you need to honestly assess your technical skills and digital know-how. As you advance on your recalculation journey, always research the following questions for any opportunity that interests you:

- What hardware, software, apps, or other technologies are most commonly used in this job or profession? If you're not sure, review the technical skill requirements on any job postings that appeal to you, ask friends in the industry, or review the LinkedIn profiles of people who have the kinds of jobs you want and note what skills, competencies, and technical qualifications they list.
- Are your technical skills and knowledge up-to-date with all of the above?
- If not, what actions can you take to reach the skill levels you need?

As discussed in Chapter Two, there are more options than ever before to access skills training. You can check out certifications offered by your local community college, or use an online source of credentialing guidance and course offerings such as Codecademy, Credential Engine, General Assembly, GitHub Learning Lab, Microsoft Learn, or Salesforce Trailhead.

One note of caution: be wary of choosing a particular certification or skill just because it is being touted as essential to obtain a "job of the future." There are countless articles about these magical jobs and skills, but nobody knows for certain what future jobs will actually arise, or can guarantee the right route to get there. It reminds me of when I was a kid in the 1980s and the prevailing wisdom was that students who wanted to succeed in life would be wise to learn Japanese, because Japan was going to be the hottest country in which to do business in the future. Little did we know that Japan was on the cusp of a "lost decade" of economic stagnation. While it is critical to keep up with the technology of today, don't get too caught up in trying to predict what you will need to know tomorrow. "Start with the market now," advises Michelle Horton of Wake Forest University, when it comes to upskilling or reskilling. "That means what you *know* to be true, not what you guess will be true."

If you are thinking about reskilling to a "hot" field, such as computer programming, and you want to avoid getting caught up in the image of getting rich overnight at a tech start-up, Tom Ogletree of General Assembly suggests dipping a toe in the actual work before fully committing. He recommends checking out free training resources as a first step, even if, for example, they're just short videos on YouTube. "Make sure you genuinely enjoy the work," he says. "Explore it, touch it, feel it in a way that is low risk before investing time or money."

What if you consider yourself a Luddite and you don't want to learn any new skills or how to use any new digital tools because you're comfortable with your current methods? If you're feeling stuck in your ways and resistant to change, remember the importance of a growth mindset once again. You can get better at anything if you put in the effort. And then, if you're *still* reluctant, it might be time to accept a difficult truth: in order to have a career today, you have no choice but to adapt to the speed of technological change and address any weaknesses in your technological competence. In the harsh but true words of retired U.S. Army general and former U.S. Secretary of Veterans Affairs Eric K. Shinseki, "If you dislike change, you're going to dislike irrelevance even more."

Step Three: Position Your Professional Experiences

The value you bring to any professional endeavor heavily relies on your real-world experience. You generally can't just say, "I'm reliable, organized, and a team player" and be hired as an office manager if you've never had the actual experience of working in an office.

Or can you?

I've found that it's more possible than people think to solve the classic conundrum, "How do you get a job without experience and get experience without a job?" We've already addressed the fact that college majors don't matter as much as people tend to believe. The same is true of our understanding of "work experience." In a climate where jobs, industries, companies, and societies are rapidly changing, there are many professions where almost nobody has the traditional definition of experience. Think about it: no one knew how to create website advertising until the Internet existed, or design self-driving cars until they were invented, or teach kindergartners remotely during a pandemic until they had no choice.

What this means for you as a recalculator is to widely broaden your definition of "experience." If you're applying for a new position, don't just define "experience" as a paid, full-time job working in a similar position and industry. Instead, recognize that "experience" can include volunteering, course-work, extracurricular activities, internships, professional association involvement, gig work, and, of course, transferable work from other professional roles or industries.

Here's an example. Picture Kevin, a reliable, organized team player who has zero office experience. He's a former nurse and has been a stay-at-home parent for the past five years. He wants to transition to an office manager job to have more regular working hours now that his daughter is in school full-time. He finds a job listing for an opportunity that seems ideal, but it asks for one to two years of office management experience.

What can Kevin do? First, he needs to control what he can and focus on the rest of the job listing. What specific tasks are mentioned? What personal qualities are requested? What is the organization's culture described or known as? Then, Kevin's objective in his cover letter, résumé, and hopefully interview is to explain his career story in a way that demonstrates how he meets all the other needs of the job and why he will be an asset to the organization—not despite, but because of his unique set of experiences.

For example:
- If the job listing mentions that the office manager needs to be calm under pressure, Kevin can tout his experience working as an emergency room nurse and handling literal life-and-death situations.
- If the job listing mentions that the office manager will coordinate the company's social activities, Kevin can describe the event plans he created as a chairperson for his daughter's preschool fundraisers.

- If the job listing mentions that the office manager will oversee employees' annual enrollment in health insurance and benefits, Kevin can explain that his former roles as a stay-at-home parent and a healthcare professional give him particular insight into the diverse needs of the company's employees and their families.

While you can only work with the true facts of your experience, here's the goal of crafting a career story to meet the needs of a particular job description: you want to make the compelling case that even though you don't meet the position's exact qualifications, you are actually an even *better* choice because of how singular your skills and experience are.

You can take this approach for any aspect of your story that you feel self-conscious about, including where you did or didn't go to school, how many times you've changed jobs, why you started a business and are now seeking full-time work, or anything else. Your narrative is yours to craft in a positive and meaningful way for the particular opportunity you are seeking. Remember recalculator rule #1: embrace creativity.

What if you really do need additional experience to land the job you want? I'll bet in most cases it's less than you think. A team of researchers recently analyzed labor market data collected from 2010 to 2020 in order to demonstrate how work experience in one profession can count as experience in another field. The findings showed that "displaced workers [i.e., those who have been laid off due to an organization's closure or downsizing] do not have to start from scratch," said head researcher Michelle R. Weise, Ph.D.

In a study of more than 10,000 displaced retail workers,

534 people were able to transition to a job in human resources by adding a few complementary skills to their existing experience, such as talent sourcing and payroll administration. Meanwhile, 328 people moved into accounting and finance by "layering on" skills in risk analysis and compliance. In another study of 6,813 displaced hospitality and food service workers, 344 people were able to move into marketing jobs by developing skills in fundraising, event management, and relationship building.

These results show that "lane changing," as the researchers defined it, from one career to another is completely possible, and movement in certain directions is more possible than others. Yes, you might need to take a class or two to gain additional skills, but you don't have to go back to school for years to obtain another degree or start at the bottom in another industry. In other words, you might be closer to a career transition than you think. The study pointed out that a retail worker might be 80 percent of the way toward a role in human resources. Or a server might be 30 percent of the way toward an in-demand role as a networking analyst.

How can you determine how closely your current experience might translate into working in another field? While Weise and her team applied sophisticated data analysis, you can do some analysis of your own—on Google. There is a trove of data to be found simply by entering "Most common career changes for _____" into the search engine. I tried this for teachers, nurses, salespeople, truck drivers, lawyers, and stay-at-home parents like Kevin, and I found plenty of different options, lists, and resources for people transitioning out of each profession.

Jessica Leigh Dow, a career strategist at the Office of Academic Advising and Career Strategy at Penn State University, shared the story of an alumnus who had graduated more than twenty years before and reached out to his alma mater's career center for support. He had worked as a line chef, then a sous-chef, and then had his own food business. He was going through a major life transition and decided, "I can't do this anymore," but he had no clue what he wanted to do.

"I told him it sounded like he had a lot of experience with supply chains in the restaurant industry," said Jessica, "and Lehigh Valley, Pennsylvania, where we live, is a major artery for logistics. He didn't want to go back to school for a full master's degree, so I suggested he consider an online graduate certificate program in supply-chain management. He went for it and a year later landed a great job in logistics."

Exercise: Cast the Widest Net Possible

Taking into account your soft skills, hard skills, and experiences, start to list all of the job functions, industries, projects, courses, and career-related activities that could be part of your recalculation. In *Getting from College to Career*, I referred to this as creating a Really Big List, and it's a great brainstorming technique for any stage of your career.

What are ALL the ways you can advance your career right now toward your desired destination based on the experiences you've had and your unique career story? This might include a combination of full-time job titles, part-time job opportunities, freelancing, entrepreneurship, or anything else you can think of. Don't limit yourself in any way. Just get all of these potential ideas out of your head and onto a document, using the grid below, a blank page from a journal or notebook, a Google Doc, a note in your phone, or any other format that will be easy for you to access and add to whenever you have a new idea. And, as it's likely you'll find yourself in a transition again at some point, the work you do now can guide you in the future as well.

Write down each soft skill, hard skill, and experience in your professional or personal history, then brainstorm all of the potential career options related to that skill or experience. Again, there are no wrong or outrageous answers, and don't worry right now about whether you would need an additional degree or credentials; just try to come up with as comprehensive a list as you can. Internet searches and group texts to friends asking for ideas are strongly encouraged!

Existing Soft Skills, Hard Skills, and/or Experience	Potential Related Opportunities
After-school babysitter	Full-time nanny, day-care-center owner or employee, teacher, child psychologist, social worker, guidance counselor, toy store or children's retail owner or employee, tutor, managing a babysitting service or website, working at an after-school program, working at a museum on children's programming, working at a child-focused nonprofit, writer for parenting or childcare website

Your Career Story, Part II: Communicating Your Personal Brand

Up to this point, I've been encouraging you to expand your thinking as widely as possible regarding all of the potential paths you can pursue based on your unique combination of soft skills, hard skills, and experiences. Now it's time to put all of these pieces together into a presentable package that you can communicate to others. That package includes your résumé, cover letters, and social media profiles, with a specific focus on LinkedIn.

As you work through the tips below, keep in mind that these representations of you need to be digestible by two very different audiences: human beings (networking contacts, recruiters, hiring managers, potential clients or customers) and computer algorithms (search engines and, if you're applying for positions with large corporations, online Applicant Tracking Systems, or ATS). I'll address both needs.

Remember that all of these documents and profiles are continuous works in progress. You can make edits at any time. You can change your mind about what you want to share. You can reframe your experience every time you make a recalculation, whether it's small or large. These are your personal marketing materials and they can and should be customized for each audience you're trying to appeal to, just as a salesperson customizes a pitch deck to each particular client.

I do this myself all the time. If I'm promoting a new book, I'll move my experience as an author to the forefront of my LinkedIn and social media bios. When I joined a new speakers' bureau, I did the same thing with my keynote speaking experience. When I thought I might want to obtain more work as a social media influencer, I highlighted certain experiences in my profile and added in more key words to position myself for that type of work. When I decided not to pursue that path, I undid those changes. You have the power and the permission to make any changes you want, as often as you want.

As you work through the steps to communicate your personal brand and career story in various formats, also pay attention to the tone or mood you are conveying, implicitly or explicitly. When you describe your professional moves, and especially your current period of recalculation, do you portray an active journey of making decisions and taking advantage of opportunities, or do you talk about "stumbling into" positions or "being the victim" of various circumstances or "chugging along" in one role or another? Do you smile or smirk when you tell your story? Yes, many people are recalculating these days because they *have to*, but you can still position any transition as a choice you're actively making and that you have a positive attitude about.

Résumé Tips for Recalculators

As I'm sure you've guessed, the vast majority of recalculations require some work on your résumé. Consider this document to be the written embodiment of your personal brand and career story; it's the "leave-behind" after a networking interaction or job interview that actually *does* go into the decision-making room as a representation of you. This is why it's important for your résumé to represent you and your career story in the most persuasive, positive way possible. Here are my top tips:

1. Don't Reinvent the Wheel.

Maybe you've just graduated from college and this is your very first time putting together a résumé. Maybe you're changing careers and your résumé needs a top-to-bottom overhaul to refocus on your new goal. Maybe you've been working at the same company for twenty-five years and haven't needed a résumé since the original *Beverly Hills 90210* was on the air.

Whatever your situation, you never need to start from scratch. It's easy to access helpful résumé templates. Do some Googling, borrow a résumé-writing guidebook from the library, or ask your university career center for examples that are similar to your specific situation. Then ~~steal~~ borrow the best ideas for layout, formatting, headings, chronology, and more.

2. Tailor Your Résumé to Each Opportunity.

You'll want to have a master template of your résumé, but it's important to customize it each time you apply for a job. Each time you send out a copy, adjust it to fit the particular opportunity by adding in different key words or rearranging the order of the bullet points under each job to match the key words, qualifications, and requested experience stated in the job description. Just be careful to keep track of which person has received which document, and note that if you're submitting your résumé to a large corporation or submitting to a college's résumé "drop" that is shared with multiple employers, you will have to choose just one version.

3. Be Mindful of Your Contact Information.

Your contact info is part of your personal brand. This includes:

Name

If your name is often mispronounced or much of your work experience took place under a different name, consider including the phonetic spelling of your name or a previously used name in parentheses. (LinkedIn allows you to record an audio of your name pronunciation to add to your profile if you'd like.) Many professionals today also include preferred gender pronouns (e.g., they/them/theirs, she/her/hers, he/him/his). If you have a unisex name and you want to identify your sex, you can include "Mr." or "Ms." in parentheses as well.

Names can be a fraught issue, especially for BIPOC. Minda Harts, author of *The Memo: What Women of Color Need to Know to Secure a Seat at the Table*, has shared that her given name, Yassminda, proved a challenge for her primarily white teachers and college professors, so she shortened it to Minda. "When I graduated college, I automatically knew I had to go with Minda for the jobs that I really wanted because that was the first thing that they would see," she has said. "We know that sometimes those unconscious or conscious biases will prevent people from picking up the phone and calling you for an interview. So it just made life so much easier at the time, I thought. But what I realized is if a place isn't going to accept you for what your full name is, then that's going to be the least of your worries going down there."

Trust your gut when it comes to your name, and, as Minda says so eloquently, don't compromise on your identity for a job that will likely not be worth it.

Contact information

For your email address, keep it as simple as possible (e.g., your full name or first initial and last name) and use a standard, current email provider like Gmail or Yahoo—AOL and Hotmail email addresses appear old-fashioned. If you're a student or recent grad, you can use your university-provided email account if you'd prefer. Avoid cutesy email addresses for sure.

For your physical address, if you live in one place but also spend some time in another (for example, if you're a college student who goes home when school is out), then you can list a current location and a "permanent" or "alternate" location. Check out the sidebar for guidance on handling the location issue when you are seeking remote opportunities. You can also note on your résumé and cover letter if you are open to a relocation.

I know it might seem dated, but make sure any phone number listed on your résumé is set up for voice mail. Many millennials and Gen Zs in particular tell me that they don't like voice mail or have never really used it, but that isn't okay if you are job hunting. Some recruiters and networking connections still leave voice-mail messages, so make sure your outgoing message is professional and clearly states your name. Just a simple, "Hi, this is Lindsey Pollak. Please leave a message" is fine. And remember to check your voice mail daily. Recalculating is challenging enough; don't miss out on a good opportunity just because you don't like voice mail and missed an important message.

Social media profile links

As a demonstration of your professionalism and tech savvy, you should always list your LinkedIn profile as part of the contact information on your résumé. (Be sure to customize the URL to be easily readable; mine is linkedin.com/in/lindseypollak to keep things simple.) On that same note, if you want to land a new position in certain industries (especially a social media company!), consider including the social media handles that have currency in that field or organization as well. More on that topic in the final section of this chapter.

Quick Recalculation:
How Can I Position Myself for Fully Remote Opportunities?

For some people who had never done so before the pandemic, working from home was a revelation. The volume of job searches using the "remote" filter on LinkedIn has increased by about 60 percent since March 2020. Some companies, notably Twitter, will not require employees ever to return to offices, although this policy is not widespread.

What can you do if you want to recalculate to a fully remote work life? The good news is that it is more possible than ever before to live anywhere and still find work that fulfills you. (Best example: the gorgeous islands of Bermuda and Barbados both responded to the pandemic by creating one-year residency visas for professionals or students who wanted to "work remotely from paradise.")

I posed some questions I've received on this topic to the leading expert on working flexibly and remotely, Sara Sutton, the CEO of FlexJobs. Here are her answers:

Q: How do I position my location on my résumé and LinkedIn profile if I'm a remote worker and/or would like to be considered for remote work opportunities?

A: Because most remote employers still have location requirements for their remote jobs (which can happen for many reasons, including state tax requirements), they'll want to know your location when you apply. Our career coaches at FlexJobs typically recommend that people put both their location and their desire for remote work on résumés and LinkedIn profiles. For example, where you've got your contact information at the top, you can put, "Boston, MA, and Remote Work."

Q: Do I just come out and say "I want remote work" on my résumé and LinkedIn or should I be more subtle?
A: If remote work is your ultimate goal and you're not considering jobs that won't let you work remotely, then it makes sense to state that clearly in your job search materials. If you're open to both in-office and remote roles, you may want to be more subtle.

Q: What advice do you have on including remote work or remote internships on my résumé and LinkedIn profile? Do I even need to mention a job was done remotely?
A: If your goal is to land a remote job, you should absolutely include information about your previous remote experience throughout your résumé. Here are some ideas from our career coaching team:

- In your job or internship descriptions, you should note when a job was partially or fully remote. Where you'd typically put the company's city and state, you can also include the level of remote work. Depending on how often you worked remotely, you might write, "Some Remote Work," "Mostly Remote Work," or "100% Remote Work."
- Job descriptions on your résumé can also discuss how you've worked across time zones or collaborated with teams of people via email, phone, and online programs even if you worked in an office.
- If you've done any remote study or education, note that in your Education section.

- A Technology section on your résumé is a chance for you to list all the communication and collaboration tools you're familiar with that remote teams rely on, such as Web and videoconferencing, online chat, and file collaboration and sharing.

- And lastly, if you've been working remotely since the pandemic lockdowns began, that can be noted as well. Something like, "Continue to be productive and effective while working 100% remotely since March 2020."

For more helpful résumé and cover letter templates and other resources for securing remote or flexible work, visit flex jobs.com.

4. Feature a Summary or Overview Statement.

If you're new to the job market or changing careers, the person reading your résumé might be unclear on exactly what job you want. This is a problem because most recruiters tell me that if they are not sure what a person's career goal is from their résumé, that document usually ends up deleted or in the "circular filing cabinet." Since "objective statements" have become outdated, adding a quick synopsis of your career story at the top of your résumé, right below your contact information, will solve this problem and ensure your goal is clear.

Here is an example of a summary statement from the résumé of a longtime advertising sales executive who wanted to make a career transition to a nonprofit or socially responsible organization:

Accomplished executive looking to make an impact at a dynamic, mission-driven organization. Leverages 20 years of revenue-generation strategy and sales experience stewarding relationships and identifying mutually beneficial opportunities. Confident and consultative partner with unwavering persistence.

The résumé featuring this summary statement helped the executive successfully recalculate from a career in the declining industry of advertising sales to the growing field of renewable energy sales. If you are pursuing multiple career options, remember the above advice to have multiple versions of your résumé and, therefore, multiple versions of your summary statement.

5. Bulk Up on Key Words.

You can supplement your summary statement with a brief "Areas of Expertise" or "Skills and Qualifications" section. This serves to highlight the details of your career story that you want your résumé reader to have top-of-mind before moving down to review your education and experience.

This is a good place to feature a list of key words that demonstrate how your skills and experience match what an employer is looking for. Review online job listings for the kinds of positions you're interested in, along with LinkedIn profiles of people who have the positions you want. Then feature the same words—if they're accurate, of course—throughout your résumé. Key words are critical to helping your résumé get through search engine and ATS filters. And, again, you can alter this section for different versions of your résumé as needed.

6. Quantify Everything That's Quantifiable.

"Managed a team of tour guides" is less impressive than "Managed a team of 12 full-time and 5 part-time tour guides serving more than 300 customers per week." If your job involved increasing profits or satisfaction, or saving money or time, feature those figures as well: "Improved call center productivity by 20% in six months" or "Raised $50,000 through dedicated email marketing campaign to targeted donors." Quantifying shows your measurable contribution to an employer while simultaneously demonstrating that you're a person who understands the importance of quantifying results.

7. Include Layoffs and Canceled Job Offers Related to the COVID-19 Pandemic.

According to Dawn Carter, a longtime recruiter who is currently the director of global university recruiting at Uber, you should absolutely include a COVID-19–related layoff, canceled internship, or rescinded offer on your résumé—as well as on your LinkedIn profile. This demonstrates that you succeeded in the recruitment process and the cancellation of the job was through no fault of your own. Be sure to use the word "canceled" as opposed to "rescinded" because the latter could imply that you personally did something that resulted in the loss of the offer. The listing on your résumé would look something like this:

XYZ Company—Summer 2020 internship offer
accepted. Canceled due to COVID-19.

If you were laid off or unemployed because of COVID-19, don't worry too much about a gap in experience on your résumé for that period of time. Dan Black, global leader of talent attraction and acquisition at EY, says, "There will be a lot of forgiveness among recruiters for a period of unemployment related to the pandemic and the resulting recession. There is no need to hide the fact that you were among the tens of millions of people affected."

Quick Recalculation: How Do I Address Gaps in My Experience?

What if you have gaps in your work experience that are un-related to the pandemic? Perhaps it's because you took a few years off to be a caretaker, or for your own health reasons, or because you experienced a prolonged period of unemploy-ment. The reason doesn't matter. Here are three steps to po-sition any break in a positive light.

1. Don't try to hide it: No matter how fancy you try to get in disguising dates of employment, a savvy person read-ing your résumé will notice. The best approach is to leave the gap on your résumé and address it in your cover let-ter with a short explanation, such as, "I took a few years off while my kids were young, and now they are back in school and I am excited to return to full-time work" or "I had been dealing with some health challenges that are now under control."

2. Highlight any experience you gained while out of the work-force: Many recalculators have accomplished some profes-sionally related activities while on a break—whether by taking a class, volunteering, keeping up with industry news, or hold-ing down a side gig. Or you might have been keeping one foot in the career door by freelancing or consulting. Make sure to highlight your commitment to continuous learning and any relevant transferable skills you picked up, like mastering a new software program, participating in continuing education courses to keep a credential (such as an attorney maintaining a license to practice law), or increasing volunteer activities at a nonprofit. If you are currently on a break and haven't done any of these things, you might want to consider engaging in a few career-related activities before you start applying for jobs, so you can strengthen your abilities, refresh your skills, and enhance your career story.

3. Reiterate your desire to recalculate: Explaining what you did during a career break is the perfect segue to explaining your recalculating goals. Come up with a short explanation of what you learned, observed, or tried while you were unemployed that helped you arrive at your new career aspiration. Remember that people love stories, so make yours compelling and then remind them that you are rejuvenated, energized, and ready to tackle a new role. With a little creativity and positivity, you can turn a gap into a nonissue.

---

8. Prioritize.

When you list bullet points under each job on your résumé, you don't have to list accomplishments chronologically. Instead, list the most important accomplishment or responsibility first. It's unlikely that a potential employer is going to read every bullet point under every item, but most people will read the first or second bullet point on each list, so always put the most impressive ones at the top.

9. Include Unpaid and Part-Time Work, Volunteering, Consulting, Internships, Extracurriculars, Projects, and Gig Work in Your Experience Section.

I've mentioned it before but it bears repeating: just because you didn't get paid or a job wasn't a full-time salaried position doesn't mean it's not work experience. I'm not a fan of made-up job titles like "Domestic Leader," but it is perfectly appropriate to include "Full-Time Parent and Homemaker" to explain how you spent a period of time and any skill- or experience-building activities you undertook during that time.

The same goes for volunteer positions and extracurricular activities for students: if you raised money, managed people, analyzed data, edited newsletters, or planned events, all of this can and should be included as work experience. For projects and consulting engagements, remember that the work doesn't have to be paid or long-term to be relevant to an employer and worthy of inclusion, and no one needs to know how many hours you worked or how much you earned (or didn't). All that matters is that it built your experience and credentials.

10. Minimize Experience You Don't Want to Promote.

As you've probably noticed, there are many ways to draw a reader's attention to what you want to emphasize, so avoid these strategies if you'd prefer to downplay any information. If a job included working on a task or responsibility that you hated and never want to do again, then don't include that bullet point at all. You can even leave off early-career and short-term jobs if they aren't relevant to your current career goals.

Another reason to minimize certain experiences is the unfortunate reality of age bias in the hiring process. Although the law forbids discrimination based on age, recruiters and hiring managers can have unconscious bias against older job applicants who have many decades of experiences listed on their résumés. If you are concerned about potential ageism, Kamara Toffolo, a résumé and job search strategist, advises that you pick key points from your early career and create an "Early Career Highlights" section on your résumé without listing the exact years each job took place.

If you're concerned about having stayed in one position for "too long" and appearing to have stagnated, Toffolo recommends that you demonstrate some degree of growth and progression in your role or organization. "Show you didn't do the same things for ten years. Show there were variations in the roles," she advises. You can accomplish this by listing the responsibilities you gained over time and the specific outcomes you achieved along the way. You don't necessarily have to show a straight upward progression—remember to embrace those creative career shapes from Chapter Two—but you do want to demonstrate development and dynamism.

11. Edit Your Education.

Although you need a section to cover your education, you can include as much or as little information as you believe will serve your goals. If you're concerned about ageism, you are under no obligation to list your college graduation year on your résumé. If your college major has nothing to do with your career path, you don't need to list it. And as mentioned in the previous chapter, your GPA is also not required or necessary.

You should, of course, list any continuing education courses or professional certifications to demonstrate that you are committed to continuous learning and/or upskilling or reskilling.

12. Don't Ever Lie or Stretch the Truth.

While I've offered a lot of strategies to highlight certain aspects of your career and minimize others, do not ever lie or exaggerate. Dishonesty is always a bad idea and almost always gets uncovered eventually.

13. Keep It to One Page.

Unless you are applying to academic jobs where long résumés (usually called a CV, or curriculum vitae) are the norm, your résumé should be one page. Over the decades I've heard all the arguments for a longer résumé, and I will never waver on this advice. The reason I am so adamant is this: your résumé is a marketing tool first and foremost. The whole point is to communicate your career story and personal brand in a concise and digestible way.

A résumé is not meant to be a transcript or a comprehensive list of everything you've ever done every day of your career. By keeping your résumé to one page, you're demonstrating that you can edit yourself and succinctly demonstrate your worth, both of which are important skills in the fast-moving professional world of today.

14. Don't Get Fancy.

In almost all circumstances, the only acceptable way to present your résumé is on a standard letter-size document in a standard black font (Times New Roman, Arial, etc.) on a white background. No colored paper, no graphics, no scent, no fancy fonts, no bells and whistles. Recruiters, especially those in the corporate world, laugh at these attempts to stand out and immediately throw such résumés away. I also recommend saving and sending your résumé as a PDF to make sure your formatting looks the same on all computers (and to ensure you haven't left any revision marks).

Above all, always follow the exact formatting guidelines of any employer's website you submit to. If they tell you to use orange paper that smells like tangerines, then you can ignore my advice.

15. Seek Input.

I don't recommend sending a résumé to a potential employer until a few other people have reviewed it for content, spelling, grammar, visual presentation, and overall clarity. If you're willing to invest a bit of money, it's often worth the cost to hire a professional résumé writer who has experience in the industry you're job hunting in, especially if you're changing careers or returning to work after time off. Another good option is to seek help from your college career center. As I've mentioned, most will provide free résumé writing counseling to alumni no matter how long ago you graduated. A final option is to ask a professional you know in your desired industry to review your résumé and offer any feedback or suggestions.

The most important feedback to seek on your résumé is whether or not you have made it clear what kind of opportunities you want. This is why I'll promote the importance of a summary statement one final time: your résumé must be *absolutely clear* about *exactly* what you want. Whenever you share your résumé with someone, try the ten-second test. (In my first book, I advocated for thirty seconds, but that feels too long nowadays.) If after reviewing your résumé for ten seconds the person can't explain what opportunities you want, then it's time to go back to the drawing board.

Craft Effective Cover Letters

Will anyone actually read this?

It's the question I asked myself in my very first job search as I carefully printed out each meticulously drafted letter, and it's a question job seekers still ask as they submit their missives through online application portals.

I ask recruiters all the time how they feel about cover letters, and the answers vary widely, from "Never read 'em" to "They're the most important factor I consider." Since you'll never know the exact weight a cover letter will have in the hiring process for a particular opportunity, you have to put in your best effort on each one. No matter what, the cover letter does give you a chance to shine in a way your résumé doesn't. It allows you to elaborate on aspects of your experience that are specific to the job requirements, and it gives you a place to positively frame a career recalculation. Here are some guidelines:

1. Think of Your Cover Letter as the Appetizer.
A cover letter should be compelling enough to entice a recruiter or hiring manager to want to know more about you.

This means it's okay to show some personality. I'm not suggesting you toss out your professionalism and use slang or emoticons (please don't!), but you can take the formal tone down a notch and make your cover letter more conversational. And don't be afraid to show some enthusiasm!

Picture the person you're addressing and pretend you're sitting across the table. You wouldn't say, "It was with great excitement that I read about your available role." No; you'd probably say, "I've wanted to work for TG Sneakers since I set a high school volleyball record while wearing a pair. When I saw that you were looking for a product marketing manager with my experience, I knew I would be the ideal candidate to meet and exceed your sales goals." Hit your experience high notes with some flair in your letter, then let your résumé do the heavy lifting with more specific details.

2. Customize the Cover Letter to the Job.

Just as I advised you to have multiple versions of your résumé customized to each job, it's even more important to personalize each cover letter. Specifically refer to the name of the company, department or team, and the exact position you want. Research the employer thoroughly (more on how to conduct employer research in Chapter Five) to match your personal attributes and career story to its culture and business goals. If you learn that the recruiter attended the same college you did, the cover letter is the place to work it in. Do you have any former freelance clients in common with the team you're applying to join? A little name dropping can get them nodding their heads.

3. Show, Don't Tell.

If the job description calls for an efficient team player, don't just write in your cover letter, "I'm an efficient team player." Instead, make those key words come alive by offering an example of how exactly you acted like a team player in your most recent job or in a volunteer role. You can share a story of how you led a cross-functional team that spearheaded an effort to streamline your company's new business presentation process or how you recruited a team of multigenerational professionals to fundraise for a local charity. Specific examples are more powerful and memorable.

4. Address Issues. Briefly.

If you've had a gap in your work history or are trying to change industries, remember that the cover letter is the ideal place to address the issue and communicate your transferable skills concisely and positively. For example, if the job posting mentions experience in management for a manufacturing company, and you have management experience in a finance organization, explain how these responsibilities are comparable, such as cultivating your team, managing a P&L, and presenting to the C-suite.

If you are a recent college grad who had an internship canceled or a job offer rescinded, you can use your cover letter to discuss how you used a period of unemployment productively, perhaps by working for a family member's small business or taking an online course to build a new skill. If you stayed out of the workforce to care for children or aging relatives, explain the break in your work experience with a single sentence. Remember: do not apologize or overexplain; just acknowledge the facts of your situation and move on.

If you are applying for a job despite the fact that you don't have the requested experience, the cover letter is also the place to make your case. Explain what you are doing to up-skill or reskill yourself—such as taking LinkedIn Learning courses or volunteering your time to build skills in your desired job function—and, again, the transferable skills you can offer.

5. Send the Letter Directly to Someone in the Company.

If you haven't been on the job hunt for many years, you might be surprised at the lack of personal touch involved in the process today. Often, you'll submit your application materials directly on a website where it will likely go through an Applicant Tracking System, which uses artificial intelligence to sift through résumés for specific skills and experience. If you don't fit the mold of what they're looking for, your application might not ever be read by a person.

This is why it's more important than ever to also send your cover letter and résumé to someone specific. As we'll address in detail in the next chapter, scour your LinkedIn connections to see if there's anyone who can help identify a viable contact, or conduct your own search on LinkedIn and the company website. Even if it's not the "right" person, they can pass it along and you'll have a better chance of your materials eventually ending up in the hands of the person making the interviewing decisions. "To whom it may concern" should be your last option.

6. Proofread, Proofread, Proofread.

You likely have read your own cover letters so many times that your brain automatically fills in words you think are there. (My friend Cathie is a former PR account executive who used to read a lot of cover letters from students who wanted to be interns at her *pubic* relations firm. Except they would miss a letter. Oops.) Another common mistake is to use the same letter for multiple employers and forget to change every mention of the organization you're applying to. Don't write a letter to Coca-Cola telling them how excited you are to work for Pepsi.

Don't rely on spell check or Grammarly to catch errors. If you have access to a printer, print out a hard copy of every cover letter to review, or try reading each letter out loud or backward to catch any errors. Send it to a friend to take a pass as well. While an error might not count you out, the cleaner your letter, the better the impression you make.

LinkedIn Profile Tips for Recalculators

From 2009 to 2015 I served as an official ambassador for LinkedIn. In that consulting role, I developed and led global webinars for job seekers, veterans transitioning out of the military and into the civilian workforce, and college career services professionals who were training students in how to use LinkedIn. I am still a huge advocate of the network and use it daily.

Like any social media site, LinkedIn is constantly adding

and removing various bells and whistles, but the fundamental ways to optimize your profile have remained remarkably consistent over the past ten years. Here is my very best LinkedIn profile advice to help you best position your profile for your recalculation journey (you'll find more on using LinkedIn for networking and job hunting in the coming chapters).

1. Turn Off Your Profile-Update-Sharing Settings.

Before making any updates to your LinkedIn profile, go into your privacy settings and turn off the automatic notifications of profile updates. You can do this by scrolling down to "Visibility of your LinkedIn activity" and toggling off "Share job changes, education changes, and work anniversaries from profile." This will ensure that the people in your network don't see a constant stream of updates on every comma you change on your profile. (This is especially important if you are currently employed and don't want your employer to notice that you are updating your profile.)

2. Write an Information-Packed Profile Headline.

When it comes to both appearing in LinkedIn search rankings and catching the eye of actual humans, the headline is the most important section of your profile. I'd actually argue that your LinkedIn headline is the most important professional real estate you have on the Internet.

Because people, especially recruiters and hiring managers, are super busy and have short attention spans, you want to offer as much information as possible and as quickly as possible. I always work from the assumption that most people don't scroll past the top "box" of most LinkedIn profiles, so I suggest putting all the good information there.

With this level of importance in mind, imagine that you're going to tweet your ideal personal brand or career story synopsis; that's how you should approach your headline. Tell people what you are doing now, what you want to do, and what key skills or experiences you offer.

When you're making a transition, always start with what you want, not what you've done in the past. If your heading says you're a college senior majoring in psychology, a recruiter will never know that you want a job at a nonprofit. You can use commas, vertical lines, or slash marks to separate the various pieces of information you want to share. The goal is to provide as much information as possible in case a person does not read any further than your headline.

Here are some examples:

- Aspiring Financial Analyst | BA in Economics | Former Intern at ABC Bank | Experience as Campus Government Leader | Fluent in Spanish
- Enthusiastic start-up marketing professional | Former high school English teacher | Currently enrolled in digital marketing certificate program
- Seeking Full-Time Position as a Web Developer | Full Stack Engineer | Front End Specialist | HTML5, CSS3, Bootstrap, jQuery, AJAX, MySQL, PHP
- Executive experienced in strategic planning, program evaluation, cost analysis, budget management, and board governance | MBA | Nonprofit board member

3. Post an On-Brand Professional Photo.

The essential accompaniment to a thorough headline is an on-brand profile photo. LinkedIn employees describe the profile photo as a "virtual handshake." If it's going to be the first time a potential networking contact or employer sees you, you want to make the best first impression, just as you would want to do in person.

My advice is to post a photo of yourself looking like you're on your way to a professional networking event or job interview in your desired industry. (This means no wedding or prom photos with your significant other cropped out.)

If you want to work in a formal environment, then you should be wearing a suit in your photo. If you want to work as a video game designer, you can dress more casually. If you are totally committed to a particular industry, then you can add context to your photo. For example, I knew a career changer who wanted to go into sports management, so for his LinkedIn profile he took a photo of himself wearing a suit inside a football stadium.

I'm not as concerned about the background photo (the wide image behind your profile picture). If your recalculation involves launching a freelance career or small business, you'll want to use an image of your logo, your products, or a representation of your services (like a photograph of yourself on-stage if you are a professional speaker like me). Otherwise, if you're looking for full-time employment, keep it simple with the standard LinkedIn background or something safe like the skyline of your city or a plain color. Don't let any fancy imagery distract from your headline and headshot.

4. Update Your Location.

Despite the remarkable growth in remote work since the onset of the pandemic, many (though not all) employers still want employees to live in the same city as their physical location. Therefore, it is wise to preemptively change your profile location to reflect where you intend to live and work, which may not necessarily be where you currently are. (Check out the earlier sidebar for advice on seeking remote work.)

College students and recent grads often overlook this detail. You might be graduating from a school in Corvallis, Oregon, or Bloomington, Indiana, but that doesn't mean you plan to live there. Recruiters often screen out applicants by location, and if you aren't a match, you likely won't be considered for the job.

5. Mark Yourself as #OpentoWork.

If you are actively job hunting, LinkedIn offers the ability to mark your profile as "#OpentoWork," either as a publicly viewable branded circle surrounding your photo or as a private badge for recruiters if you want to be discreet about your job search. The company launched this new feature in the months following the COVID-19 outbreak, and it was quickly adopted by job seekers at all levels and across industries. Don't be shy about using this feature.

As my former clients Omar Garriott and Jeremy Schifeling, who now run a business called The LinkedIn Guys, advise, "You never know who will see your LinkedIn posts about your job search and forward your information to someone they know. A longtime family friend might know a business owner who needs a marketing assistant. Your friend's girlfriend from high school could be hiring a software developer." In the first few months that #OpentoWork was available, LinkedIn reported that members of the site with the photo frame received 40 percent more InMails (private messages) from recruiters and 20 percent more messages from the LinkedIn community.

6. Tell Your Career Story in the About Section.

Use your profile's About section to show recruiters why you'd be a great addition to their team. This is not the place for a dry list of past job titles, especially if you are seeking a full-time position. It's also a missed opportunity as a recalculator to leave this section blank. Basically, you want to answer the question, "Who are you and why should I hire you?" It's essentially a cover letter and sales pitch, and I recommend writing in the first-person perspective.

The opening sentence is the most important. Don't be gimmicky and don't tell your life story from the very beginning. Cut to the chase. You can use words like "aspiring," "future," and "passionate" to indicate your level of experience. Make the case for the kinds of jobs you'll be applying to or, if you're seeking consulting work or self-employment instead of—or in addition to—a full-time job, the work you'd like to be offered. Mention relevant projects that tie to specific skills, even if you worked in a different industry or role, or even if all of your experience has been in internships, volunteer roles, or extracurriculars. Don't go into detail about experiences that don't relate to your current career goals.

As with your cover letter, your About section is a good place to express your enthusiasm about the pivot you're making. It's also appropriate to use this space to mention a layoff or canceled opportunity related to the pandemic, as long as you keep a positive tone. Ada Yu, group product manager of the team that helps people find jobs on LinkedIn, also recommends adding into your About section that you are willing to work remotely if that's the case.

7. Play to Your Strengths or Weaknesses as a Writer.

There is no law that says the About section of your profile has to be written in full sentences and paragraphs. If you're a strong writer, then by all means show off that skill. If you're not, write a sentence or two, then switch to bullet points that highlight the top skills and experiences you want people to keep in mind as they scroll through the rest of your profile.

No matter what, it's important to speak in your authentic voice. Otherwise, when people connect with the real you, it will feel inconsistent with the "you" they read about on LinkedIn. For example, I once had a webinar attendee ask if it was okay to be funny in his LinkedIn About section. My honest answer was, "It depends. Are you funny?" If you are, then go for it! If you're not, don't try to use humor as a strategy to get a job. Always be authentic.

8. Add Relevant Sections.

Over the years, LinkedIn has added sections that you can use to boost your profile, but this is not an invitation to make your profile as long as a CVS receipt. Be very selective about which areas you opt to include. Here is some guidance on which sections to consider adding to your profile:

- **Education:** This is essential but, as on your résumé, you can choose what to include and what to leave out.

- **Licenses and Certifications:** Include any credentials that are relevant to your current career goals, as well as widely applicable and well-respected credentials like a CPA. This is also the place to show off any reskilling or upskilling you've been working on as part of your recalculation, including the completion of LinkedIn Learning or other courses.

- **Volunteer Experience, Organizations:** Definitely include either or both of these sections if you are an active volunteer or participant in extracurricular activities or professional associations, even remotely. It's a sign of your commitment to your community and the causes you care about. Note, however, that if you participated in a leadership role in any of these organizations and are lacking paid work experience, then listing these roles in the main Experience section of your LinkedIn profile might be the better option. Remember that it's up to you how you categorize your various activities.

- **Skills and Endorsements:** This is the place to feature key words that align with your personal brand and the opportunities you're seeking as a recalculator. Go ahead and delete any skills from your profile that you don't want to use any longer. For example, if you want to pivot from a people management role to an individual contributor role, then delete skills such as "supervising others" from your list. LinkedIn's official recommendation is to choose at least five skills to feature on your profile. I think that's a good guideline, but I wouldn't list more than ten. You also might add skills related to remote work such as "working with and managing distributed teams," suggests Ada Yu. And don't worry too much about how many people "endorse" you for each skill. "Pin" the top skills you want to be known for at the top of the list so that they are visible on your profile, regardless of whether they have received the most endorsements from other people. What's most important is that these key words can be viewed by others.

- **Publications, Patents, Honors, and Awards:** By all means include these if you have them and they are relevant to your career goals. If you have multiple achievements (bravo!), I'd limit your profile to the top three to five representative examples. If you've received an award or honor that is not clear from its name—e.g., the Jane Doe Award—be sure to explain what achievement the award is celebrating.

- **Courses, Projects:** These are important categories for students and recent grads in particular. Even if you have little to no work experience in a certain realm, you can promote any relevant courses you've taken or projects you've worked on that relate to the opportunities you are now pursuing. Don't list every course you are taking or have taken; just feature the ones that are relevant to your current career goals. In the early years of your career, you can potentially make up for a lack of paid work experience by demonstrating your passion for a topic by posting a paper you've written, a slide deck you've created, or a video of a presentation you've given.

- **Languages:** I'd recommend only listing a language if your skill level is, at the very least, conversational. Here's the test: If, during a job interview, the hiring manager started speaking to you in this language and you could comfortably respond, then list it on your profile. Otherwise, don't include it.

9. When in Doubt, Leave It Out.

Less is more when it comes to LinkedIn profiles. Remember, people are busy and have short attention spans these days, so no one is going to read your profile word for word, top to bottom. For this reason, I challenge you to be as concise as possible (except in your headline, which is limited by character count anyway).

Imagine that I charged you $100 for every word on your profile. What words would be worth buying? Reduce the number of experiences you list, the number of bullet points under each experience, and the number of additional sections on your profile. The less information you provide, the more likely someone will actually read it.

10. Request Recommendations.

Less is more when it comes to recommendations, too. You'll want at least two or three, but I don't see any advantage to having dozens. No one will read more than a few, and recruiters tell me that if they want to know other people's opinions of an applicant, they'll ask for formal references.

For any recommendations you *do* feature, it's perfectly appropriate to ask your recommenders to use specific words or examples that reflect your personal brand and the career story you want to tell. Dawn Carter of Uber even recommends to those who have been laid off that you ask your former boss to write a recommendation—as long as you have maintained a good relationship and the layoff was through a downsizing. Dawn herself has written LinkedIn recommendations for employees that were impacted in recent layoffs. Knowing the mindset of her fellow recruiters, she not only touts her former employees' positive qualities but also outlines the reasons other employers should hire them.

Finally, as with your résumé and cover letters, have a few people review your LinkedIn profile for clarity, grammar, spelling, visual presentation, authentic voice, and, again, whether it clearly conveys your current career goal.

Quick Recalculation: How Do I Use LinkedIn If I Am Undecided or Have Multiple Career Goals?

Many recalculators want to, and should, keep their options open. If you are undecided or are exploring multiple potential career paths, I would recommend focusing your LinkedIn headline and About statements on your personal brand and transferable skills. Using your About summary as a cover letter, you can write generally about how you want to contribute to an organization's success or use your skills to support others.

Ada Yu of LinkedIn suggests highlighting transferable skills that work across industries. You can mention these in your headline, and sections like About, Experience, and Skills & Endorsements. "The most in-demand skills across industries right now are communication, problem solving, and project management," Ada says, adding that a LinkedIn Learning course can help you brush up on these skills if needed. "Once you finish the course, add the skills to your profile and highlight the certificate of completion in the Certifications section."

If you're pursuing both full-time employment and entrepreneurial or freelance work, another option is to use your LinkedIn profile page for job hunting and then create a separate LinkedIn company page (similar to a Facebook fan page) for your business. You can certainly list your business on your main LinkedIn profile, but separating the two will keep people from becoming confused.

Social Media Tips for Recalculators

While having a LinkedIn profile is considered appropriate, valuable, and reliable across almost every industry, there are many other social networks that can support your recalculation. To decide whether or not it would be helpful to use other platforms for your particular situation and goals, first ask yourself these two questions:

1. Which Social Network Presences Are Valued in Your Desired Industry?

If you're not sure how to answer this question, do a Google search of some of the top professionals in your dream industry. Which social media platforms are they active on? You can ask recruiters you meet at job fairs or virtual networking events which social media profiles are commonly listed on résumés in the field you're interested in. You can also ask contacts in that field whether they use social media for professional networking or career development.

Frankly, many professions do not require you to have a social media presence in order to be successful. You don't have to be on Instagram if you want to be an insurance broker, or on Twitter if you want to be a social worker. In many industries, there is little downside to avoiding social media.

On the other hand, if you want to work in fashion, beauty, or the art world, you'll find that many professionals in those fields are very active on Instagram. If you want to work in political communications or journalism, Twitter is often a must. Keep in mind that, when more people began working remotely during the COVID-19 pandemic, some professionals became more active on social media than they had been in the past, because it was the only option for professional networking. You might be surprised by who is active and on which sites, so do your homework and don't make assumptions.

2. Which Social Networks Do You Personally Like to Post and Engage On?

If you already enjoy and spend time on a certain platform, then that's a good indication that it could be an asset to your recalculation. Many Class of 2020 college graduates who had job offers rescinded posted about their circumstances on sites like Facebook and Instagram, along with a request to their networks for help. With one simple post, they began crowdsourcing for jobs, tapping into a community that already knew them and wanted to be supportive.

This kind of social media engagement worked because it was authentic. I wouldn't recommend launching a Facebook account just to tell people you're job hunting. What I do recommend is pivoting accounts on sites where you are already active to be more career-focused.

If you do plan to use an existing social media profile for

job searching, personal brand building, and other career development purposes, I would suggest that you first give your profile a once-over. Are there any embarrassing or inappropriate photos or posts—or even ones that just project a negative or snarky attitude—that you might want to remove so networking contacts or potential employers don't see them? Are there political or sports tirades that might not reflect the personal brand you want to project? Have you posted anything negative about a past employer or client? These decisions highly depend on your desired career path (that sports rant could be an asset if you want to be a sports commentator), so they are truly up to you.

Overall, keep in mind that, unless you set your accounts to "private," any social media presence you have may be viewed by a professional contact at any time. It is called the World Wide Web, after all. It's worth taking some time to Google yourself and review any profiles you have out there in the world to make sure you're comfortable with anything discoverable about you.

If you have an account on a platform that you no longer use, my recommendation is to just delete it. You want to avoid a situation where someone reaches out to you with an opportunity or networking introduction and you never get the message. If you don't want any professional contacts to see a particular social media profile, then go ahead and make it private.

Christine Cruzvergara, vice president of higher education and student success at Handshake, a website for entry-level jobs and internships, also advises job seekers to consider your "digital shadow," which includes what other people have posted about you on social media. Recruiters, hiring managers, potential clients, and networking contacts may see this when they Google your name, which they often do. "Remember that nothing is really private," she says.

Have honest conversations with friends and family if there are posts you'd like them to take down or screenshots of text messages you'd like them to delete. Err on the side of caution. It's a miserable experience to lose an opportunity because of a tweet or text, even one posted by someone else, and unfortunately it happens more often than you may think.

Turning back to the positive, if you believe your accounts will contribute to your recalculation goals, then now is the time to become more active on your favorite sites. List professionally relevant profiles on your résumé. Post more frequently about industry issues. Update your social media bios to reflect your personal brand attributes and the headline key words you used on LinkedIn.

When it comes to what you're actually sharing and posting on your social media accounts, again consider the career story and personal brand you want to project. If you write in your cover letters that you are passionate about the music business, then people will expect your social media postings to include posts about music, articles shared about music, and photos or videos from music performances.

You should also be following industry leaders and employers in the field you are applying to work for. This demonstrates genuine interest and a desire to learn and connect. "Make sure that how you engage across various social platforms is consistent with the brand you are trying to convey to employers," says Christine Cruzvergara. "And remember that what you choose to like, retweet, share, and comment on from other people says just as much about you as what you yourself post."

Here is how social media activity can lead to real job opportunities, a topic we'll continue to explore in the next two chapters. Owen was a business school student in Philadelphia who knew he wanted to transition from management consulting to the very competitive field of private equity (PE). He spent a lot of time following New York City–based PE firms on Twitter, along with professionals who worked for those firms. He followed the companies that those firms invested in. He also followed the reporters who wrote about PE.

For a few months, Owen just observed his Twitter feed and continued to follow more people and organizations. Which topics seemed to draw the most discussion and debate? Which firms seemed to be most respected by the others? Whose voices stood out? After he felt a bit more comfortable with the subject matter, he began to retweet the people he admired. He would occasionally reply to comments and people would occasionally respond back to him. He would share his perspective as an MBA student and occasionally ask for the PE professionals' input on topics he was studying in an academic setting.

One day, Owen noticed that several of the people he followed were tweeting about attending an upcoming conference in New York. He reached out to the organization hosting the conference, explained that he was a student, and received a discounted ticket to attend. When he first arrived at the conference venue, he picked up his name badge, then walked up to some of the most powerful leaders in private equity and said, "Hi, I'm Owen, the MBA student you've chatted with a few times on Twitter. It's great to meet you in person." He then proceeded to start a conversation with each person about issues he knew they were interested in because they had already started the conversation online.

Not surprisingly, everyone he met was impressed with his passion for the industry and knowledge of it as an outsider. By the time he graduated from business school, Owen had received four job offers from private equity firms. He accepted one of them and, several years later, he is now head of business development for one of the start-ups that his firm invested in. And it all started simply by following a few accounts on Twitter.

Now that you've thoroughly crafted your career story and personal brand and begun to communicate it, it's time to expand your audience. In the next chapter, we'll address how to widen and diversify your network so you can continue moving closer to your recalculation goals.

Networking in the "New Normal"

Tap the Power of Relationships by Building, Maintaining, and Leveraging Your Contacts Both in Person and Virtually

If you want to go fast, go alone. If you
want to go far, go with others.
—AFRICAN PROVERB

The most memorable piece of advice I received at the very beginning of my career has remained relevant for twenty years and counting: keep building your contacts. No matter how much the world changes, there is little you can achieve in life without other people. That's true in the good times and especially in the challenging ones.

While many of the rules for human relationships are centuries old (treat others as you want to be treated), professional relationships since 2020 have become a little more complicated (treat others as you want to be treated . . . on Zoom). I appreciate that it can be tougher to build and maintain relationships when you're operating fully or partially remotely, but networking is a critical aspect of successful recalculating and can never fall by the wayside. Here are the reasons why:

Personal Referrals Are the Best and Fastest Way to Land a New Job.

A recruiter once told me she usually has two stacks of résumés printed out on her desk for any given position: a very tall stack of résumés that have been submitted through the Internet and a very short stack of résumés that have been sent to her by people she knows and trusts. It is difficult (but not impossible, as we'll address in the next chapter) to stand out from other job applicants when you apply to a position through a company website or job board, especially if you're a recalculator trying to transition into another industry, return to work after an absence, or begin your first meaningful job after college. It's networking that gets you into that short stack, where your résumé is much more likely to be read.

Personal referrals are golden, and the numbers support this: According to a 2016 report from the U.S. Bureau of Labor Statistics, 70 percent of jobs are found through networking. And, for those of you who are in need of a job as soon as possible, a 2018 study from HR Technologist found that referred talent gets hired 55 percent faster than those who are applying through online career sites.

Your Network Increases Your Reach.

If you're searching for opportunities alone, you will only know about the ones you discover yourself. If you tell other people what you're searching for, they can be on the lookout for opportunities to pass along to you. In some cases, you might even help others by involving them in your recalculation: many employers offer significant bonuses to employees who refer talent that ultimately gets hired.

Remember recalculator rule #5: ask for help. "Now is not the time to be shy; you have to be unapologetic about the fact that you are looking for a job or internship," says Christian Garcia, associate dean and executive director at the University of Miami's Toppel Career Center. "Most people genuinely want to help."

That's what Robin Solow, the HR executive you met in Chapter One, found as she reached out to her contacts after being laid off. At first, she admits she felt embarrassed to publicly share her layoff news, but she soon realized that wouldn't get her anywhere, so she started making calls and sending emails. "There was a natural opening to connect because of the pandemic; it felt comfortable to check in with people and say, 'What's going on with you?' and 'Here's what's going on with me.' You have to put your ego aside."

The power of networking is not just about the people you know but the people they know and so on. Robin estimates that her job hunt ultimately put her in touch with nearly one thousand people. "You can have the best résumé in the world," she says, "but it won't mean anything if you haven't made the right connections." Robin ultimately got her new job at a commercial real estate start-up through her fourth degree of separation.

Squad Support Is Powerful—and Practical.

Many of the recalculators I interviewed for this book pointed to the importance of networking with other recalculators. Jill Sammak transitioned from a corporate career to starting her own business, and she said, "The one thing I wished I had known sooner would be to surround myself with others who have made a career change, and better yet, I would have surrounded myself with people who were already pursuing or doing the job that I wanted. If we spend most of our time with people who are in the profession that we want to leave, we are less likely to see the possibility of doing something new."

I often recommend the same to people who ask me for advice as they launch careers as authors or professional speakers. When I first went off on my own, most of my friends still worked in full-time office jobs, so during the day I felt really isolated and like no one believed I was truly "working." I slowly began to attend networking events and join listservs for entrepreneurs and aspiring thought leaders, and I found a community of other people in similar circumstances to mine. These new contacts helped me to feel less alone, provided a sounding board for ideas and challenges, and gave me a community to ask questions about things like how to find disability insurance and what website hosting service was best for small businesses.

Robin Solow found a helpful network of job seekers who had also been laid off and could help each other by sharing and liking each other's LinkedIn posts, participating in virtual happy hours, and just having sympathetic friends to reach out to. In turn, she made a point to reach out to and help other job seekers in her personal network, including interns she had hired and her first boss who was furloughed at the same time. "Being able to help was also a help for me," Robin says.

It might seem counterintuitive, but fellow recalculators are your allies, not your competition. Teaming up with a network of other job seekers or consultants or entrepreneurs seeking investors not only gives you a support system but also a referral network. It's rare that you'll be seeking the exact same opportunities as someone else, so when you come across an opportunity that's not a fit for you, you can share it with another person in your network—and, of course, vice versa.

Humans Can Tell You Things the Internet Can't.

Recalculators often find themselves in new territory. You might be entering the job market for the first time in a decade, or relocating to a new city, or working remotely for the first time, or trying to expand a side hustle, or learning to code. In these types of unknown circumstances, you'll likely be doing a lot of online research to figure out how to transition successfully.

The Web is great for finding data like the best law firms for working mothers or the average salary range for entry-level insurance salespeople in Tampa. On the other hand, it isn't great at answering questions like, "Will I feel comfortable as a Latinx transgender engineer at this company?"

Online employer review sites like Glassdoor and Handshake, and industry insider sites like Above the Law (a behind-the-scenes look at the industry of law) and Wall Street Oasis (an online community for aspiring professionals looking to break into investment banking, private equity, asset management, and other corporate finance careers), can provide some insight, but for sensitive questions about culture and inclusivity, which have an enormous bearing on your success and happiness, you need to ask a human being.

Many people tell me they don't like networking because they're uncomfortable asking for help. A Gallup study found that networking is the number one job skill for which recent college graduates wish they had received more training. If you feel this way, I urge you to recall your growth mindset and tell yourself you're just not great at networking . . . *yet*. If you've ever made and kept a friend, then you know how to network. I also firmly believe, to echo Christian Garcia, that people genuinely want to help each other. Case in point: in late March and early April 2020, when the pandemic first began to spread in the United States, searches for "how can I help?" spiked on Google.

If it's the word "networking" that makes you feel like you're bothering people or conjures images of schmoozy conference attendees doling out business cards like candy, then reframe the concept as building professional relationships or making friends—or even helping a company find its next terrific candidate (i.e., you). After all, says Dan Black of EY, "I'm always looking for great talent for my firm. Wherever I find that, it is never going to be a bother for me. That's my job!"

The fact is, you will be infinitely more successful as a recalculator if you become a more active networker. Remember recalculator rule #5: ask for help. I know that many people applauded the invention of the GPS because it meant they could stop asking other people for directions, but on your career journey you still have to do it!

The flip side of asking for help is offering help to others. When you reframe networking as giving as much as asking, it can feel more comfortable. All relationships work best when they are mutually beneficial, so keep that top of mind when you're making professional connections.

This is true even if you're just starting out in your career or industry. Fran Hauser, author of *The Myth of the Nice Girl*, reminds recalculators, "You are just as valuable as someone 'big.' Often, people tend to question what value they can offer somebody who's more senior. In reality, there's always an opportunity to add value. You may have access to a network or community, knowledge of a social media platform, or functional expertise that could be very helpful to someone else."

Fran experienced this herself when a young woman, Anya, reached out to her on LinkedIn to ask for career advice. At the same time, Anya offered to share some social media ideas for the start-ups that Fran advises. The offer demonstrated not only that Anya had a generous approach to networking but also that she had done her homework, researching Fran's projects in order to make an offer of support that would truly have value.

"But I Don't Know Anyone!": Identifying Your Professional Network

I wish I had a dollar for every person who has told me in a job-hunting workshop or webinar that they can't network because they don't know anyone. This is particularly common among first-generation college students and people who have taken time off from the workforce. When I dig deeper, I usually find that what people actually mean is that they don't know any senior-level executives or people in the professional fields they're interested in.

My response to this concern is that your network begins with the people you know from your everyday life, which means that *everyone* has an existing network. While you might not personally know someone at every organization or in every industry where you want to work, you probably have a contact who can introduce you to someone in *their* network who has those connections. Yes, some people start with innate advantages, but everyone has the ability to expand their network, connection by connection.

Karen Ivy, director of career services at Texas A&M University–San Antonio, tells the story of Marco, a first-generation college student who thought he had no professional connections. His dream was to be a pilot, but he wasn't sure where to start.

"I suggested he find someone in the field, so he went to a pilots association website and found the profile of a man who was a leader in the association," Karen says. She helped Marco craft an email that introduced himself and asked questions about the pilot's long career—and it worked wonders. "This man was so impressed with the outreach that he became a mentor to Marco," she says. "They met on the phone every Sunday and talked for about an hour each time. After a few months, Marco announced to our class that this man offered to pay for the training to get his pilot's license!"

We probably can't all be as lucky as Marco, but research

has shown that we are all more connected than we know. The evolutionary psychologist Robin Dunbar, a professor at the University of Oxford, has found that while the average person makes thousands of acquaintances over their lifetime, they ultimately maintain their relationships with an average of 150 people.

So, if you know 150 people and each person in that group knows 150 people, then you are potentially one introduction away from knowing 22,500 people. Each of these individuals represents a potential connection to a career opportunity. The power of networking is about those second-degree connections and beyond—not the people you know, but the people *they* know, and so on.

Dan Black of EY is a huge proponent of those second and third degrees of connection. "Don't be afraid to network in your community and social circles," he says. "I met two job candidates over the last six months via my social channels: one was through my daughter's softball team and the other was from just walking around in my town and having a conversation. You need to put yourself in situations where you might come across people other than the usual twenty or thirty people you see all the time."

Additionally, when you start to reposition yourself as a recalculator, you might begin to interact with your existing contacts in new ways. Maybe your neighbor has also been meaning to recalculate but never mentioned it until you did. Maybe your spouse's cousin has been learning to code but didn't know you were interested in talking about it until you broached the subject.

Sometimes, believe it or not, the people closest to you don't even know they have a connection that could be helpful. A junior in college named Brittany wanted to start her career in the entertainment industry, which is notorious for needing a connection to break into. Living in a small town and attending a small college in New England, she didn't know anyone in New York or Hollywood. After attending one of my workshops, she took my advice to join LinkedIn, but she was too intimidated to connect with anyone but her mom, who worked for a small business in their town.

After Brittany's mom accepted her request, she suggested that Brittany scroll through her LinkedIn connections in case she found anyone who might be helpful. Brittany was skeptical; she and her mom talked every day, and her mom had worked at the same small importing business for decades. Did her mom have some secret celebrity connection she had never mentioned?

It turns out she kind of did. When Brittany scrolled through the people her mother was connected to on LinkedIn, she came across a woman, Yael, who was a former producer for a major syndicated talk show.

"You're kidding!" her mom replied when Brittany called to tell her. "I went to high school with her. We reconnected last summer at our thirtieth reunion, but I only knew she was a stay-at-home mom. I had no idea what she did before that."

Brittany's mom sent an email introducing the two to each other, and Yael happily agreed to a networking call with Brittany. On that call, Brittany impressed the former producer with her passion for the entertainment business. Yael put in a good word and passed Brittany's résumé along to her former colleagues. Brittany landed an internship at the show shortly thereafter.

In this story, you can replace "mom" with a former high school teacher, niece, coworker, neighbor, college academic adviser, or anyone else in your life. Even the most basic networking action with someone you already know can lead to a real opportunity.

Exercise: You Know More People Than You Think

Take some time to identify the people you know. Open up a Word doc or a notes app or grab a pen and paper, then challenge yourself to write the longest list you can come up with. Take a look at your phone contacts, your social media connections, and your old address books. The goal is not to judge how "connected" or "helpful" any of these people might be to you; just make a big list. Consider including:

- Family members
- Friends
- Neighbors
- People you frequently interact with through social media sites, listservs, discussion boards, or video games
- Parents of your child(ren)'s friends
- Your friends' parents and siblings
- Current and former colleagues from every job, internship, and/or volunteering position you've ever had
- Former classmates, teachers, professors, guidance counselors, sports coaches

- Fellow members of religious organizations
- Fellow members of professional associations, charitable organizations, fraternities, or sororities
- Local business owners
- People you regularly run into at the gym, grocery store, community center, hair salon, or elsewhere

Next, make a separate list of all the organizations and institutions you're affiliated with. Consider people who share these affiliations with you to be members of your extended network. They are more likely than a complete stranger to answer your email, take your call, or accept your LinkedIn request if you mention the connection. Consider including:

- Towns or neighborhoods you've lived in
- Schools that you or anyone in your immediate family have attended
- Organizations where you've worked full- and part-time
- Clubs, volunteer organizations, professional associations, fraternities, or sororities that you've been a member of
- Religious institutions where you've worshipped

It doesn't matter if you ended up listing 15 people and organizations, or 150, or 1,500. What matters is that you have a starting point. Try sharing your lists with a significant other, parent, or close friend; maybe they'll remind you of people or organizations you may have left off. Then, keep this list handy as you work through the rest of this chapter and book.

The SPEC Approach: How to Network as a Recalculator

I absolutely love meeting people and building relationships for the sake of community and human connection, and I hope you do, too. But let's be honest. You are likely reading this book because you are in a period of career transition at a complicated and challenging time in workplace history. You are networking in this environment for the very specific purpose of finding a new job, starting a business, or changing roles in your existing organization. That requires a very targeted type of professional networking that I refer to as SPEC: Specific, Proactive, Enthusiastic, and Consistent.

Specific

Job seekers in a recession tend to say things to their contacts like "I'll take anything" or "Let me know if you hear of anyone who's hiring." This is also a common approach taken by recent college grads or veterans transitioning out of the military into civilian careers. I completely appreciate the attempt to be flexible, have a "can-do" attitude, and remain open to any opportunities. The problem is this approach just doesn't work. It's too broad and unmemorable.

If, instead, you tell people in your network that you are seeking marketing or PR opportunities with a small business or tech start-up in the Phoenix area, now you've given them some facts to grab on to and keep in mind as they go about their days. They're more likely to notice these potential opportunities and mention them to you.

Being specific in your networking requests also demonstrates to others that you have done the work of figuring out what you want to do. This is why we spent so much time in the previous chapters working on your mindset, preparation, and personal brand: clarity and self-knowledge don't just benefit you; they signal to others that you have your act together and are ready to excel at any opportunity they refer to you.

Even if you really would take any job, or you're casting a wide net and searching for opportunities in numerous industries, try to be specific and focused in your networking conversations. You might reveal to a few trusted sources that you are exploring many career options, but not every person needs to know about all the opportunities you are pursuing. This is especially true if you are connecting with someone who works at one of the organizations where you'd like to land a job.

For example, one of my closest friends, Courtney, has a senior marketing job at a major TV network. She is often approached by students from her alma mater for advice about how to break into the television industry. At one point, Courtney was approached by two students, and she generously made time to speak to both of them. They were both impressive, but one student talked all about her passion for TV and the shows on Courtney's network in particular. The other student revealed her indecision about pursuing a career in television, film, or digital media. Courtney enjoyed networking with both students, and she had a genuine desire to help them both. But when an entry-level opportunity opened up, which student do you think she referred to the hiring manager? The student who was focused on a career in television, of course.

Again, you don't have to force yourself into a single career path before you're ready, and you can share your uncertainty with people you trust. But when it comes to recruiters, hiring managers, and senior professionals, your networking efforts will be far more likely to yield real opportunities if you appear to be focused on one specific goal.

Proactive

Networking is always a two-way street. However, when you're the one who needs opportunities, the burden is on you to do most of the outreach and follow-up work. Being a proactive networker includes:

- Reaching out to connect with people you know on LinkedIn instead of waiting for them to connect with you.
- Sending individual emails to people in your network to let them know you are in the midst of a career recalculation instead of hoping they come across your announcement on Facebook.
- Approaching people at networking events and introducing yourself instead of hanging out by the food table and hoping they will approach you.
- Signing up for one-on-one time with employers at virtual career fairs instead of wondering if your uploaded résumé will get noticed.
- Following up on introductions people make for you instead of expecting the other person to reach out first.

I know these actions are uncomfortable for some people, perhaps if you're an introvert, a mid-career returner, an industry newbie, or a first-generation college graduate. For those in the latter group, you might draw inspiration from my good friend Fred Burke, who was a first-generation college grad himself and is now the director of the graduate career management center at Baruch College's Zicklin School of Business in New York City.

Fred worked throughout school to pay for his undergraduate education at the College of Saint Rose in Albany, New York, and credits the school's student affairs staff with encouraging him to network. "They coached me and made introductions and taught me professional etiquette," Fred shares.

Now that he is a student-affairs professional himself, Fred goes out of his way to demystify job hunting and career development. He advises all of his students, but especially those who aren't as comfortable with networking, to be forthright about what they don't know. For example, toward the end of college Fred decided he wanted to go to graduate school but didn't know anyone who had an advanced degree. He decided to walk into the graduate admissions office at his college and ask, "What do I need to do to go to grad school?"

"Be genuine and honest when you don't know something," Fred advises. "Never pretend you're something you're not. Then, when you ask for advice in a genuine way and show that you really want to learn, people will respond." All you have to do is make the first move.

Fred's advice is equally applicable to recalculators with decades of experience and success. Author and digital brand strategist Peter Thomson says that each time he makes a significant change in his life—including leaving law for design, moving internationally, and writing a book—he has coffee with fifty people to ask for their advice on his plans.

According to Thomson, "The hidden insight in the fifty-coffees idea is that the biggest changes in your life will only happen through the people that you meet and conversations you have. . . . If you change the conversations that you're a part of, then your life changes automatically."

Quick Recalculation: What If People Reject My Networking Efforts?

Fear of rejection is a normal and understandable emotion. It can be especially strong if you've recently experienced a career disappointment, like a demotion or a layoff, and you just don't want to suffer through more.

One of the ways to handle this fear is to remind yourself not to take things too personally. That's the advice of Kevin Grubb, executive director of the career center and assistant vice provost for professional development at Villanova University. He points out that a person's lack of response or delay in responding likely has little to do with you, especially in today's tumultuous times. "Give others space and grace. If they don't respond to you, recognize that they may be going through a lot of change and turmoil, too."

Also keep in mind the wisdom of career guru Steve Dalton, a career programming director at Duke University's Fuqua School of Business and author of *The 2-Hour Job Search* and *The Job Closer*. He says, "A lot of people will say no or ignore your networking outreach. That's fine, because the people who do say yes will be inordinately helpful to you." Steve calls these yes-people "boosters" and defines them as the ones who "are an additional set of eyes and ears for you. They empathize with you. They will give you ideas and help you navigate their personal networks. They can think of things you didn't even fathom." Steve admits these boosters will be just a small percentage of those you attempt to contact, but that's okay. "You've got to kiss a lot of frogs," he says, "but the princes are worth it."

Enthusiastic

Positivity is really appealing, so don't be afraid to show passion and excitement when you're networking. Tell people that you're energized about transitioning to a new career path. Explain why you love the work you're planning to do, the freelance career you're going to launch, the new skill you're in the process of learning, the leap you're making from retirement back into the workplace, or the new town you want to move to. Your attitude will influence how other people perceive you. If you're really excited about your recalculation, those around you can't help but feel the same way.

Consistent

The final key to successfully networking as a recalculator is to be consistent in your approach. This comprises two types of consistency: the frequency with which you reach out to people in your network and how you go about doing it.

When it comes to frequency, I'm not going to be overly prescriptive. Many job search books and experts offer detailed schedules and spreadsheets of how often you should reach out to networking connections. Their advice is valid, and I encourage you to find and use these resources if you'd like. My message on consistency is simple: don't fall off the face of the earth. I can't tell you how many times someone has contacted me to set up a phone or video call to talk about a job search or career transition, I respond, and then they disappear. Or we schedule a call and they don't show up. Or we exchange email or LinkedIn messages for three weeks straight, and then they don't respond to my latest message for three months.

I believe that inconsistency happens because people feel really motivated at the beginning of a recalculation but then lose steam. Or, in the case of the networkers who "ghosted" me, maybe they realized I wasn't the best person to talk to and they didn't know how to opt out of our call or conversation. Or they just changed their minds. If any of this is the case for you in a networking situation, all you have to do is say something like, "I'm all set for now, but I really appreciate your willingness to help."

In terms of consistency of style, remember that the actions you take when networking are also reflections of your personal brand. As the Zen proverb says, "The way you do one thing is the way you do everything." Each email, phone call, text, Zoom, and LinkedIn connection gives people information about you. If you are very polished and professional in your first networking interaction with someone, and then overly casual and unengaged in your follow-up, the person won't know which version is the real you, and they certainly won't recommend you for an opportunity. On the other hand, if their interactions with you are consistent, they will feel confident in touting your qualities to others.

Slack, Snapchat, and Starbucks: Where Networking Happens Now

Now let's take your SPEC approach and apply it to the different environments where you can find and nurture professional relationships. In the past, "professional networking" often meant putting on your best business attire, entering a beige-carpeted hotel ballroom and slapping on a name tag, then trying to make conversation while balancing a paper plate of veggies and dip in one hand and a room-temperature beverage in the other.

Sometimes networking still involves some or all of the above, but our globally connected and increasingly virtual world now offers a larger variety of opportunities through technology. This is good news if you're more of an introvert or if you have other personal obligations or physical limitations that make it difficult to attend live events. It's also good news if you're a recalculator who is not yet established in your desired industry, location, or profession because most online networking provides a lower barrier to entry than being accepted to a country club or professional society. It's also safer when, you know, there's a global pandemic.

Steve Dalton of Duke thinks the experience of COVID-19 has fundamentally changed most people's approach to networking—in a positive way. "The pandemic called upon us to act and learn in different ways and be at home more with a lot more isolation," Steve says. "When working remotely I miss my coworkers, as a lot of people do. That has made us all more open to connecting with strangers. As odd as it sounds, I'd rather be job hunting now, in this time period, than in the recession that started in 2008. I think we're all in a more humanistic place now and that will last a long time."

I agree with Steve's positive assessment, and I've been really encouraged by the generous networking I've observed since the pandemic began. When people ask, "How are you?" they really want to know the answer. When people say, "Let me know if you need anything," there's a deeper sincerity behind the offer. But you have to make the first move to tap into all of this goodwill.

In this section we'll consider the many virtual and IRL (in real life) places you can establish or reestablish professional networking relationships. Do keep in mind that in none of these venues is it appropriate to directly ask someone for a job. The point of networking is to form relationships, then set the stage for continued conversations. A request for direct help with your recalculation would happen later down the line. Right now you're just seeking a first date, not a marriage.

LinkedIn

In the previous chapter you polished up your LinkedIn profile. Now let's use it for networking outreach. Don't be shy about doing so—this is the entire reason LinkedIn exists! I recommend connecting with everyone you know on LinkedIn by working through the list you created earlier in this chapter.

Note that your connection requests are more likely to be accepted if you customize each one with a brief note rather than simply clicking "Connect." Here are a variety of approaches to that connection-request language based on your relationship to the person in question:

- For someone you're in touch with frequently in the "real world," such as a close friend or family member, just write a short message, something like, "Glad to see you on LinkedIn—let's connect!" or "Would you connect with me here on LinkedIn to support my career transition?" This is to overcome any awkwardness of connecting professionally when you are personal friends. A generic request with no message can be perceived as too impersonal by someone who knows you well. Is a personalized note 100 percent required for someone you know well? No. Is it still a nice gesture that you should make anyway? Yes.

- For someone you've met through school, an internship, a part-time job, volunteer work, or another shared experience—especially if the person was senior to you—you might consider reminding them how you are connected. For example, "It was a pleasure to volunteer with you on the Montez campaign a few years ago. I'd love to stay in touch." Or "I was an intern at ABC Inc. last summer and would be grateful to join your network here on LinkedIn."

- For someone you don't know personally but with whom you share an affiliation, especially a college, university, or grad school, try something like this: "As a fellow State U. alum, I'd love to connect. I hope you'll consider joining my network." One of the best strategies for finding classmates and other alums is the LinkedIn Alumni tool. Find your alma mater's page on LinkedIn and click on the "Alumni" link. You'll be able to search your fellow graduates by their class year, current location, employer, industry, and even top skills. You can also search by key word. This tool allows you to search for fellow alumni in your dream position, industry, or company. They might be able to share some insights about building a career like theirs, or even introduce you to the right people if you notice a great position open up at their company.

There is no statute of limitations on how long ago you knew someone in order to reach out and connect on LinkedIn. Nor is it in any way inappropriate to request a connect with anyone who has shared a club, military branch or specialty, professional affiliation, or even elementary school with you. If someone mentions that they went to my alma mater, worked for a company that I've done business with, or grew up in my hometown of Norwalk, Connecticut—or even that they have the name Lindsey/Lindsay/Lyndsey/Lyndsay and write "great name!" in their request (seriously, this happens more often than you think)—then I'll almost always accept.

- For someone you don't know and have no affiliation with, such as a person you admire from afar, you can still reach out on LinkedIn. For "cold" outreach like this, you have to demonstrate that you're making the effort to connect in a meaningful way and provide some reason for the person to say yes. Take a quick scan through their profile so you can refer to something relevant. You can mention that you admire someone's career path, or gained a lot of value from their TEDx talk, or that you read and enjoyed an article or book they wrote (hint, hint—that's the way to my heart). When you genuinely admire someone—or their work—all you need to do is tell them so in a polite, specific, and succinct way. For instance, "Rob, I'm an avid listener of your podcast and would love to connect with you on LinkedIn to follow more of your work. Thank you for adding to my knowledge as an aspiring pharma professional."

The worst thing that can happen is someone denies or ignores your connection request, and that's not a very big deal. No one will ever think, "How dare you reach out to me on LinkedIn to network professionally!" That's the whole point of the network and people understand that.

Once someone accepts your request, go ahead and capitalize on the moment. You can send a brief follow-up note inviting them to connect by email or phone, or just telling them a bit about your recalculation and asking for their advice. Remember to be specific and enthusiastic.

Facebook, Instagram, Twitter, and Snapchat

While these social media sites are obviously not designed for professional networking like LinkedIn is, they can be surprisingly effective places to establish or reestablish personal connections that turn out to be professionally helpful. In fact, one of the macro workplace changes over the past twenty years has been the increased commingling of "professional" and "personal" networking, thanks in part to the increasing ubiquity of social media. Aside from your Facebook or Instagram newsfeed, where else can you find your boss, your yoga instructor, your college roommate, your high school best friend's brother, and your great-aunt all in one place? For Gen Xers, baby boomers, and traditionalists this may feel a little strange, but millennials and Gen Zs are often more accustomed to a world where people interact with their personal and professional "friends" all in one environment.

If you want to professionally connect with people on a not-exactly-professional site or app, try following the person (if you're not already), reacting to their posts, and then reaching out in a casual way with a direct message (DM). Then you can build up to having a professional conversation in a different medium. For instance, if you follow a former colleague on Instagram, you can respond to a story she posts about baking bread by praising her sourdough skills, then suggesting a catch-up call or coffee meeting. That follow-up call or meeting would then be the appropriate time to talk about your professional goals.

The same scenario can happen on Facebook, Twitter, Snapchat, or any other social network. This is similar to how you might chitchat with a neighbor on the sidelines of your grandchild's Saturday morning soccer game, or run into an acquaintance at Target and suggest meeting up soon to continue the conversation. I wouldn't suggest DMing anyone your résumé via social media, but I think it's always appropriate to casually touch base in this environment and then suggest further communication by phone, email, video call, or in person.

Social media sites are also good places to cultivate new connections. I frequently follow professionals I admire, especially on Twitter and Instagram, to learn more about them and engage in their online communities. If there are people you respect in your field and they are active on social media, be sure to follow them, like and reshare their posts, and leave positive comments. After a while, you can choose to reach out through a DM or email and introduce yourself as someone who has interacted with them and their posts—just as I recommended above for "cold" connecting on LinkedIn. If the people do a little research, they'll notice your efforts to engage on social media and may be more likely to reply to your networking request in a positive way.

If you're looking for more of a community on social media, consider networking through LinkedIn groups and Facebook groups, some of which are private and require a certain membership and others that are open to the public. You can also participate in Twitter chats, which are group discussions on the network that take place at a regularly scheduled date and time where all participants mark their tweets with a particular hashtag. Check out the website Tweetreports.com for a running schedule of chats on a wide variety of topics.

One caveat: your social media presence and interactions have to be authentic and well intentioned. I wouldn't recommend joining Instagram just so you can follow people and network with them professionally, much like you (hopefully) wouldn't wander around the grocery store every weekend just hoping to "bump into" a specific business contact. If your presence and participation on a particular site aren't genuine, people will likely not respond in a positive way—or at all—to any outreach, personal or professional.

Webinars

Webinars provide fantastic (and often free) opportunities to learn about different topics and develop new skills. Many people don't realize they are a terrific backdrop for networking as well. I picked up this tip from my friend Chelsea C. Williams, CEO of College Code LLC, a resource for talent development and retention. Chelsea recommends that re-calculators use the online platform Eventbrite to search for webinars where you can both learn and network. Think of key words that have to do with your career interests, tack on the words "free webinar," then do a search on Eventbrite and check out what comes up. "Location doesn't matter for a virtual event," she says, "so you can attend any webinar taking place anywhere in the world."

Chelsea practices what she preaches. Right when the pandemic started, she signed up for and participated in many webinars on various topics in her field of human resources and career coaching and found that many people in the chat area were sharing that they were looking for jobs, seeking clients for their businesses, or in career transition. "I was amazed by how webinars brought so many people together," she says.

In these sidebar chats, Chelsea would briefly introduce herself with one sentence about her business, share her LinkedIn profile URL, and reach out after each webinar to others who shared similar information. Chelsea received multiple paid speaking and consulting inquiries from people she met using this approach.

I love this strategy, because on a webinar, everyone's interested in the same topic, and that gives you something to talk about when you connect. Plus, there is little to no risk. If you say hello and share your LinkedIn profile, and no one responds, so what? I now introduce myself in the chat of any webinar I participate in, and similar to Chelsea, this has resulted in several new connections and business opportunities.

Volunteer Activities

Volunteer experiences are another low-barrier-to-entry way to meet people who share your interests. Volunteering can take place in person or online, and there are infinite organizations, events, and activities. If you're looking for ways to get involved, try searching through sites like VolunteerMatch and Idealist.org to find opportunities that reflect the issues you are passionate about.

While the main purpose of volunteering is to do good for the world, it is also appropriate to network, as long as you are authentically building relationships and not aggressively trolling for job offers or clients. You can also use volunteer activities to strengthen and showcase the very skills and qualities you are trying to build or promote as part of your recalculation, such as leadership, teamwork, writing, accounting, and event planning. Many organizations are happy to assign you to a role that matches your career and networking goals. This can help you build your résumé while also introducing you to new people.

Membership and Industry Organizations

If you already have an affiliation—such as a professional association, school or company alumni association, sorority, fraternity, or club—now is the time to become a more active participant and to reposition yourself with your recalculation in mind. While it's wise to join new groups related to any potential career paths you are considering, don't overlook affiliations from your previous industry or industries. Recalculating has become so common—49 percent of people report that they've made a major career change at some point—that it is not unusual to network within a field you are transitioning out of. Some organizations even let you participate in events or online forums regardless of whether you're a paying member or not.

This was a strategy used by Jessica Scott, a licensed professional counselor and career coach, who remained an active networker while taking six years out of the workforce to care for her ailing mother and then her two children. She stayed involved in her professional organization and attended trainings for continuing education credits to maintain her professional credentials. It was always a little bit awkward, she admits. "People's name badges say what they do, so I was like 'Hi! Just getting my CEUs [continuing education units]!'" But she used those trainings as a chance to meet new people and maintain connections with others in her field. When she reentered the workforce and was looking for jobs, those connections were invaluable.

Here are some other ways to network through any organizations you are already a member of or can join to enhance your recalculation efforts:

- **Member directory profiles:** Update any existing profiles with a sentence about your new career path. For example, on a mechanical engineering professional association website, you can begin your bio by saying, "I'm a former mechanical engineer currently transitioning to a career in financial services." You want your information, including all contact information, to be current if you connect with anyone in the association and they decide to look you up on the site. (If you are quietly job hunting, of course you can ignore this tip. But it's always smart to review any profiles you have in any forum to make sure they reflect the personal image you want to project. I can't tell you how often I come across people's bios and email addresses that are years or even decades out of date.)

- **Newsletters and social media:** Sign up for e-newsletters and follow each organization on the social media platform(s) of your choice so you don't miss any announcements about upcoming events and opportunities. You might even offer your skills as a writer, photographer, IT professional, marketer, and so on, to demonstrate your value to your fellow members and the association leaders, who are often in the know about career opportunities within their membership. When I was first building my business, I wrote as many articles as I could for various associations' websites, to establish name recognition and Google search results for the areas in which I wanted to build expertise.

- **Job boards:** Some organizations have job boards that are available to both members and nonmembers. If you're actively job hunting, take a look at each of your associations' websites, and if they have one, post your newly updated résumé and look through posted opportunities. Often these smaller, more focused job boards can be more effective than the larger sites. If you're seeking freelance or consulting work, many organizations also have "marketplace" sections where you can post your capabilities and members can post projects you can apply for.

- **Online forums:** Join Slack channels, LinkedIn Groups, Google Groups, Facebook Groups, and other forums. I hear a lot of resistance from recalculators about joining these types of forums if they haven't participated in the past. Complaints range from "There is *so much* posting!" to "So many discussions are totally irrelevant or uninteresting to me." I get it, but thanks to the volume of content that people share on these forums, there's probably a lot that might be *very* relevant to you, including real career opportunities. You just have to spend some time searching for the needles in the haystack. Also keep in mind that an organization's most active and dedicated members are often the ones who spend time participating in these forums, which means they are, as Steve Dalton said, disproportionately likely to be helpful.

Quick Recalculation: Networking with Recruiters and Headhunters

As you can tell, I love to help recalculators explore a wide variety of options. You know who doesn't like to do this? Recruiters and hiring managers with real jobs to fill. You should only connect with a recruiter if you are seriously planning to apply—or have already applied—for a job that this person is actively trying to fill. Otherwise, don't waste their time by reaching out about anything else, unless that person happens to be a close friend of yours.

In times of high unemployment, many recruiters want to publicly promote that their employer is hiring—it helps to boost the organization's reputation—so they will often include "I'm hiring" or "We're hiring" directly in their social media headlines. You can consider this a flashing neon sign declaring, "Reach out to me!" when you come across a position you want.

That said, recruiters are often inundated with messages from job candidates, so you do need to stand out from the crowd. Susan Rietano Davey is a former recruiter and current cofounder and owner of Prepare to Launch, LLC, a company that provides coaching to professionals in work-life transitions. She receives countless emails and LinkedIn messages from job hunters, and they're easy to overlook. But Susan has some simple and surprising advice on how to stand out in today's world: "Pick up the phone and call me!"

Since barely anyone calls anymore, job candidates will stand out by doing so. "It's harder to hang up on someone than to ignore their written message," she says. Even if job searchers can't reach her, she's more likely to follow up if they leave a friendly, well-crafted voice mail. "Your voice is more intimate; it reveals your personality better than typed words," Susan says. "Plus, we are so digitally connected now, it's a pleasure to hear a human voice."

You'll find more in-depth advice on working with recruiters in Chapter Five on acing the job search.

How to Make Every Networking Conversation (a.k.a. Informational Interview) a Success

Now let's address the heart of networking, when someone agrees to chat and you are preparing to have a one-on-one conversation with them. This is also referred to as "informational interviewing," which is just a formal name for learning about a potential career or employer by talking to people who work in that field or organization. I prefer to use the broader term "networking conversations," because these can and should take place throughout your career and not just when you're hunting for a new job. It's those "fifty cups of coffee" that Peter Thomson talked about.

Whether a networking conversation takes place at Starbucks, in the hallway at a busy conference, on a Zoom call, or via email, the goal is the same: share your career goals and ask a more experienced or knowledgeable person for advice.

There is a lot of nuance to this type of conversation for a recalculator. It's not that complicated for an accountant to go to a networking event for accountants and ask if anyone has advice on getting a job as an accountant. It's a little more complicated to be a computer science major who worked as an investment banker, then became a stay-at-home parent, and now wants to join a biotech start-up. The answer to "So, what do you do?" or "What can I do to help?" requires a little more explanation.

Here's how to maximize your chances of having a successful networking conversation:

Impress from Your First Outreach Message.

You begin to make an impression from the moment you send an email request or place a phone call. Every interaction a potential networking contact has with you will increase, or lessen, their desire to help and support you.

In most cases, I recommend making the ask for a networking conversation by email or a message on LinkedIn. When communicating by email, Debbie Epstein Henry, a public speaker, author, and expert on careers and women, advises that you use the subject line to indicate exactly why you are getting in touch. Some people receive hundreds of emails a day, so your message has to stand out to be clicked on. "By scanning the subject line of an inbox," she says, "a recipient should be enticed to open your email and also know right away what it is you're seeking." Be specific. Instead of a generic subject line like "Networking" or "Request," Debbie recommends that you include the person's name and details such as, "Laura—Fellow State U. alum seeking expertise on asset management careers," or "Jon—Connecting about a transition to nonprofit fundraising."

In the actual message, remember your SPEC approach:

- Be **specific** about the advice you'd like to receive.
- Be **proactive** about asking for a certain amount of time—I usually start by requesting fifteen to thirty minutes and offering multiple options for connecting (e.g., by Zoom, phone, or a coffee chat at a location convenient to them).
- Be **enthusiastic** about your desire to connect with this person.
- Be **consistent** with the personal brand you are presenting in relation to your LinkedIn profile or any other information they might find about you through a Google search. I recommend including a link to the social media profile that best portrays who you are—usually LinkedIn. However, do not attach a résumé or any other documents. Most people will not open an attachment from someone they don't know, and it is too forward to attach a résumé before someone has agreed to speak with you.

Here is an example of an outreach email message:

Subject: Fellow Feldman campaign alum—connect about environmental consulting?

Hi, Anne,

I was a volunteer on the Feldman campaign last year and admired your work on the candidate's environmental platform. I recently left a corporate role in operations and I'm working to transition to a career in the clean energy space. I'm deeply committed to improving the environment here in California and beyond. Would you be willing to share some career advice with me, perhaps in a fifteen-minute phone or video call or a few email questions? I would be so grateful for your time and support. Thank you for considering my request.

Best regards,
Will

Own the Logistics.

As the person requesting help, it's your responsibility to schedule the phone call, video call, or meeting, create and send a calendar invite with the details, and confirm the appointment at least twelve hours in advance so the person knows for sure that you'll show up. If it's a call, confirm if the person prefers to be on camera or not—some people have very strong preferences, and no one wants to be surprised. Then, if you've scheduled a 2:00 P.M. call, be sure to call at 2:00 P.M. on the dot. This shows that you respect the person's time and that you are taking this opportunity seriously. You'll impress them even before you start talking about your career goals.

Do Your Homework.

While you should always do a bit of research before asking someone to connect, be sure to do a more comprehensive review of the person's background and career before an actual networking meeting or conversation. Take time to research the person on LinkedIn, Google, social media profiles, and anywhere else they appear prominently (e.g., their employer's website and/or their own website). What can you learn about this person from what they have shared publicly? This will help you to come up with smart questions to ask, request specific support, and ensure you don't waste their valuable time asking questions that could be answered by a basic search, like their education or employment history. Here's the difference:

> *Question asked by person who did no homework:*
> Are you involved in any charitable work?
> *Question asked by person who did extensive homework:*
> I really enjoyed your recent LinkedIn post about the volunteer coaching you do with veterans who are starting businesses. How did you get involved with that organization?

With well-researched questions like the second one, you're demonstrating that you truly want to get to know this person and their interests, rather than just asking a series of generic questions. When you get someone talking about topics they're really passionate about, you often learn the most and build the most genuine connection. I recently accepted a networking request from a student who had prepared about five questions to ask me in a thirty-minute phone call, which struck me as the perfect number. She clearly had specific issues she wanted to discuss and I appreciated the fact that she had prepared so thoroughly. Her questions were also a good mix of specific (regarding a particular interview she had coming up) and general (she asked me what books I was currently reading and would recommend to a recent grad).

Be sure to ask some questions that are not directly related to the job you want, advises Kevin Grubb of Villanova University. Otherwise, he says, "The conversation will feel transactional and could end quickly." He suggests posing a question like, "What are you finding are the biggest needs your organization has right now?"

"Not only are you demonstrating eagerness to learn about your contact's expertise, but you're also opening the conversation to go in a broader direction," he notes. "With a question like this, you also may gain specific industry insight that can help you determine skills to gain or ways to share your experiences in a future job application or formal interview."

Clearly and Concisely Explain Your Situation.

In most cases, the person you are interviewing won't know much about you (don't assume that he or she has checked out your LinkedIn profile or any other information). So it's a great idea to start the call with a brief one- to two-minute introduction to who you are and what you're looking for. This is where all of that work on your career story and personal brand comes in handy. For instance, "I've just graduated with a BA in computer science and I've completed a few internships at big companies. My goal is to find a job at a start-up in the Austin area where I can work in product development."

If you're not totally sure what you want to do, do your best to give the person some indication of the fields or roles you're interested in. For example, "I've just graduated with a degree in communications and, although I'm not completely sure yet what career to pursue, I'm currently looking at positions in public relations and marketing."

Be a Great Listener.

Listening is a wildly underrated skill, and it's a game changer for networkers. In a world when we are constantly interrupted and talked at—from advertisers and social media influencers to text message pings and buzzes from family and friends—it is a joyful experience to complete a full sentence without being interrupted. (As my grandfather used to say when I talked too much, "You have two ears and one mouth. Use them accordingly.")

Says Steve Dalton, "Job searches and career success in the modern world are not about making a sales pitch. That brings people's guards up. To network successfully, you need their guards to come down and that happens by listening."

Listening means not interrupting when you ask a networking contact for advice and they start talking. Listening means genuinely caring about the other person's story—there is wisdom to be found in the details of someone else's journey. Listening means taking notes.

Listening also provides you with clues about how to make sure every networking interaction is mutually beneficial. Always ask, "Is there anything I can do to support you?" and truly listen and digest the answer. If the other person asks you for a recommended resource, a connection to someone you know, or anything else, that should be your first action item following the networking encounter.

Ask for Assignments.

This is my all-time favorite informational interviewing tip and it works every time: ask every networking contact to recommend specific actions or assignments. You might ask them to recommend a favorite book, podcast, blog, or Twitter feed. (Pro tip: have some ideas of what you'd recommend so you can return the favor if they ask for your recommendations as well.) Perhaps ask them about a potential employer or professional association to research, a class to take, or a skill to build. The goal is to obtain actionable advice. This approach has two benefits: First, you avoid receiving vague advice ("Keep your chin up!") and ensure you have specific tactics to pursue after the meeting. Second, you have the perfect reason to follow up and keep in touch with your networking contact: to report back on taking the specific action they recommended.

If, for whatever reason, you don't want to follow up on a recommendation that person has made, that's fine. Just say something like this either in the meeting or in your follow-up message: "Thank you so much for recommending *The Life-Changing Magic of Tidying Up* while I explore the idea of becoming an interior designer. I so appreciate the suggestion and will keep it in mind as I work on next steps. I'm grateful for your time." Then, just don't read the book. If they persist and ask again if you followed up on their advice, you can say something like, "I'm currently pursuing other avenues, but thank you again for the recommendation. I'm really grateful for your help."

Follow Up with Gratitude and Action.

The best opportunities often arise in the networking follow-up process after both sides have had the chance to digest and reflect on the conversation. But the process of following up actually begins right at the end of that initial chat or meeting. Always close a networking interaction by asking the other person how he or she would like to keep in touch. With so, so, *so* many places to communicate nowadays, don't stress yourself out by trying to guess. Just ask, "What is the best way to stay in touch with you?"

You can also set the stage for follow-up based on when you expect to complete various stages of your recalculation. If you're in the early, exploratory stage, you might want to say that you'll be in touch in a month or two. If you're busy with your current job while looking for another one, you can say that you'll be in touch in a few weeks. The time frame is up to you; what's important is setting those expectations and then meeting them.

No matter what your overall time frame is, your first order of business after any networking interaction is to send the recipient a heartfelt thank-you, ideally mentioning a specific piece of advice that you found particularly helpful. The best networkers send a thank-you message within six to twelve hours. (I used to say twenty-four hours, but the world moves faster today and it doesn't take much effort to write an email or text as opposed to handwriting and mailing a card, which used to be the norm.)

Share your gratitude using the person's preferred method of communication; this shows that you were listening. And, of course, if you promised to help them in any way, do that as soon as possible. Remember, too, that if someone has agreed to a networking conversation, that person now has an investment in your success and wants to hear how you're doing as your recalculation continues. Be sure to keep each person posted, especially when you ultimately land a job or achieve your desired goal. Everyone wants to feel that they contributed, even in a small way, to your success.

Build a Personal Advisory Board

My final piece of networking advice is to build a personal advisory board consisting of your most trusted and helpful contacts. A personal advisory board is a metaphorical group of people (you won't actually have meetings) that you turn to for ongoing advice, guidance, and decision-making support.

Although I love the concept of mentoring, it can be difficult, given the constantly changing nature of careers today, to find one single Yoda-like person who can advise you, decade after decade. A personal advisory board, which you may consider in addition to a mentor if you already have one, gives you the opportunity to receive mentoring from multiple people. I encourage you to create as diverse a collection of board members as possible so that the advice you receive is different from what you might come up with on your own and inspires you to think differently. Remember recalculator rule #1: embrace creativity.

Exercise: Build Your Personal Advisory Board

List the people—the founding members of your personal advisory board—to whom you'll turn for advice and guidance on your recalculation journey and beyond. These people don't have to "accept" their nomination or even necessarily know that you consider them advisory board members. The goal is to create a diverse list of people to help support, motivate, and guide you when you need various forms of support.

Consider friends, family members, former bosses and colleagues, former teachers, professors, and coaches, among others. Consider people of different races, ethnicities, generations, sexual orientations, personality types, geographic regions, industries, and job functions. The more diverse group of people you can tap for advice and support, the more innovative your decisions and actions will be.

1. _____

_____

2. _____

_____

3. _____

_____

4. _____

_____

5. _____

_____

6. _____

_____

7. _____

_____

8. _____

_____

9. _____

_____

10. _____

_____

Once you've crafted your advisory board, take a moment to think about whose advisory boards *you* might be part of. Remember, as always, to give as much as you want to receive.

When you practice mutually beneficial networking and show others that you are genuinely invested in their success and grateful for their support of yours, people will want to continue being part of your network—no matter what direction your career journey takes you. Make a point to continue to engage in communities that support you during your recalculation, and do "small goods" for the people in your network and on your personal advisory board on a continual basis, such as liking their social media posts, attending webinars they promote, commenting on articles they write, and, of course, offering to help in their career endeavors. One of the net benefits of any recalculation is the new people and organizations you encounter that remain with you as you embark on your new direction.

5

Ace the Job Search

**Build Your Expertise in Twenty-First-Century
Job Hunting, from Job Boards to Employer
Research to Virtual Interviewing to
Embracing Imperfect Pancakes**

Every time I thought I was being rejected for something
good, I was actually being redirected to something better.
—STEVE MARABOLI

If you feel like you're not very skilled at job hunting, or you're
out of practice, or you never learned how to do it in the first
place, rest assured you are not alone. As Steve Dalton of Duke
tells his MBA students, "You've never been trained to do this!
It is unconscionable that algebra and history can be required
learning, but finding a job—which is a universal source of
anxiety and determines whether you can support yourself
and your loved ones and put food on the table—is somehow
NOT required."

Most people have to figure out how to job hunt on the fly. It's a complicated and ever-changing task, made more so with the rise in virtual methods and requirements since the onset of the pandemic. The good news is that there are many experts and resources to guide you through, and I've tapped many of them for wisdom to share with you. Even if you're not actively job hunting as part of your recalculation, the tips in this chapter can help in any situation requiring an application process, interview skills, or negotiation strategies.

Do Not Pass Go as a Job Seeker Without Visiting Your College Career Center—Even as an Alum

Sometimes I think my business as a career expert is built on my shame around never visiting the career center when I was in college. What was I thinking? Its entire purpose is to help students and alumni find and succeed in their careers—for free (well, not counting the tuition you've paid). I encourage you to take advantage of the many services that career centers provide, whether you graduated last year or fifty years ago.

"Your alma mater's career services function can be a great lifeline during any transition period you're experiencing," says Kevin Grubb of Villanova University. His top recommendation to alumni is to use the services as a way to check in and update their job search fundamentals. "Alumni might think we only help people at a certain stage of their careers," he says, "but the truth is that we're hearing and learning things about recruiting, job seeking, and the employment market that can benefit a variety of people." Some career centers even offer services to members of the local community, so be sure to research any colleges or universities in your area in addition to your own alma mater. You have nothing to lose and everything to gain.

Here are some of the services you can tap into, mostly virtually and sometimes in person, too:

One-on-One Career Coaching

A career counselor can help you at a variety of points in your recalculation: deciding to make a transition, brainstorming potential opportunities related to your goals, administering and reviewing assessment tests, discussing the best ways to reskill if necessary, and creating a networking plan.

Résumé and LinkedIn Review and Improvement

It's always valuable to have an expert review your career-related marketing materials, and most offices have counselors who specialize in different industries.

Virtual and In-Person Mock Job Interview Prep

While career center professionals have always conducted virtual interview prep with alums, they became even more adept at this when many employers switched to virtual-only recruiting in 2020. These professionals can introduce you to the latest virtual interviewing technology and help you practice your answers to the most common interview questions. I offer many tips below, but practice and personalized feedback are incredibly valuable.

Workshops, Networking Events, and Career Fairs

One of the critical functions of a career center is to educate students and alumni about all the different facets of career development, including internships, freelancing, and entrepreneurship. The centers offer frequent webinars, workshops, company information sessions, career fairs (live and virtual), and networking events for students, alumni, and employers. One of the major advantages of attending an event hosted by your alma mater is that you're guaranteed to have your college experience in common with the people you meet.

Introductions

Remember Steve Dalton's recommendation that the people who say yes to networking requests will be inordinately helpful? Career center professionals epitomize that statement, and they are particularly fond of making introductions among students, alumni, and employers. They are, in particular, extraordinarily connected and knowledgeable about the careers of people who attended your school. Don't be shy about asking them for names of people you might reach out to for an informational interview, employers who often hire alums of your school, or any other introduction you're seeking.

Exclusive Job Listings

If you're a student or recent graduate (up to five years or so), setting up job alerts and posting your résumé to your college career center's job board are absolute musts. While students may be aware of the big companies that recruit very publicly on their campuses, a lot of smaller employers post jobs as well. Many alumni will share their employer's job postings with their alma mater's career centers to avoid the mass of applications they would receive on a large, public job board. This is especially true for "just-in-time" hiring, which means jobs that become available anytime throughout the year, as opposed to during the more formal campus recruiting "seasons" of fall or spring. If you're a recent grad, continue to review listings during the summer and after you've graduated so you don't miss these openings.

While most of the jobs you'll find on career center job boards will be entry-level, there are occasionally positions for more experienced candidates, so it's always worth checking no matter how long ago you graduated.

Start Applying, Stat!

Okay, recalculator: I'm going to give you some advice that might surprise you. If you're in need of a job and haven't begun applying yet, put down this book right now and go fill out an application or submit a résumé. Just find any job that looks half-decent and go for it. Remember recalculator rule #2: prioritize action. Especially if you haven't job hunted in many years (or you've never job-hunted before), you need to get in the game and practice going through the process—yep, even before you read the rest of the advice in this chapter.

This is also the first piece of advice Christian Garcia of the University of Miami offers to job-hunting students and alumni: "When I ask job seekers how many positions they've actually applied for, the answer nine times out of ten is one, or maybe two. I tell them, 'You can't get a job if you don't actually apply!' Some people are just so scared to pull that trigger to just apply."

He stresses that this is particularly important for people who feel that they are at a disadvantage in the job market for any reason, whether they are part of an underrepresented group, an older professional fearing ageism, a returning worker with a large résumé gap, a career changer with a complicated work history, or a person with a disability.

"The deck is stacked in favor of certain people and there are a lot of inequalities that make things harder for certain people," Christian acknowledges. "That being said, you have to do the work."

Michelle Horton, the career services professional at Wake Forest University School of Business, agrees. "I had no access as a first-generation college student. What I did have is a strong work ethic. I leveraged the things in my control: hard work, perseverance, and excellence. They never go out of style."

I came across a remarkable story of hard work and persistence on the discussion boards of Wall Street Oasis, the online community for aspiring finance professionals. In a discussion titled "A Guide to Surviving the Recession for New Grads," a community member named "Neanderthal" shared the following about his quest to land a job in financial services during the 2009 financial crash:

> I sent 500 resumes. Everything in my gut told me
> to quit after the first 200 resumes. I kept going. I
> felt like an idiot after 300 resumes and plenty of

interview rejections. I still kept going. After 400, it really seemed hopeless, but I kept at it. . . . At some point, you probably should quit and change paths. However, at the same time, it seems that great career breakthroughs only happen long after you've lost hope, yet you continue trying anyway. In fact, I'd say that this is one of the reasons that there are so few financially successful people. You almost have to be irrational. The rational person quits and goes on with an average life. I can't blame them—they made the rational decision! People like me are the weirdos who don't understand "no."

Is it irrational to apply to five hundred jobs to make it into the industry you want? Maybe. But I think it's just as irrational to apply to only one or two. To paraphrase hockey legend Wayne Gretzky, you lose out on one hundred percent of the jobs you don't apply for.

Make the Most of Job Boards

Now that you've submitted at least one application, work to fill your inbox with many more potential opportunities. Start to set up email alerts on as many job boards as you can to alert you to potential matches for your skills and interests. Yes, you are more likely to find a job through networking, but people do also land jobs from posted listings on job boards, and these sites have other benefits, too, including:

- Developing a sense of the overall job market—who is hiring and for what positions
- Finding additional key words from relevant job postings that can be added to your résumé and LinkedIn profile to improve your chances of being noticed
- Discovering organizations and even industries you hadn't known about that could be a great fit for your recalculation

As a general rule, the more niche the job board, the more value you'll derive. I encourage you to research job boards specific to your region, career level, industry, location, and more. Revisit the "Cast the Widest Net Possible" exercise from Chapter Three to note all of the industries whose job boards might have opportunities for you. Here's a list to get you started:

- **General job boards:** CareerBuilder, Glassdoor, Indeed, LinkedIn, LinkUp, Monster, SimplyHired, ZipRecruiter
- **Job boards for college internships and entry-level opportunities:** College Recruiter, Handshake, Intern From Home, Parker Dewey, Scouted, Vault, WayUp
- **Job boards for workers over the age of 50:** AARP Job Board, Retirement Jobs, SeniorJobBank.org, Workforce50
- **Job boards for people of specific races, sexual orientations, and members of other underrepresented groups:** ability-JOBS (people with different abilities), AsianHires.com (Asian American), BlackJobs.com (Black), DiversityInc (all underrepresented groups), DiversityJobs (all underrepresented groups), LatPro (Latinx), OutProNet (LGBTQ+), TJobBank (transgender)

- **Job boards for women and mothers returning to the workforce:** Après, Career Contessa, The Mom Project, Women's Job List
- **Job boards for executive-level opportunities:** ExecuNet, HeadHunter, TheLadders
- **Job boards for remote opportunities:** FlexJobs, Remote.co, We Work Remotely, Working Nomads
- **Job boards for freelance work:** Bark, Fiverr, Freelancer, Guru, PeoplePerHour, Upwork
- **Industry-specific job boards:** Administration Jobs (admin), Bookjobs.com (publishing), CoolWorks.com (outdoor), CrunchBoard (tech, start-up, and engineering), Dice.com (tech), eFinancialCareers (finance), Energy Jobline (energy, oil and gas, renewable energy, offshore, and power/nuclear), Good Food Jobs (food and restaurant), HealthcareJobsite (healthcare support, technician, nursing, and physician), Idealist (nonprofit), ITJobPro (information technology), Jobs in Logistics (logistics, supply chain, distribution, transportation, warehousing, freight forwarding, manufacturing, purchasing, and inventory management), Jobs in Manufacturing (manufacturing), MarketingHire (marketing and advertising), Mediabistro (media and communications), SalesJobs.com (sales), USAJobs (federal government)

Once you've identified a few job boards to start with, here are some tips for applying through these sites:

- Triple check each job board profile you create. Every time you set up a new account, you'll be asked to fill out a lot of the same information, including contact information and work history. I know this kind of data entry can be tedious, but don't drop the ball on accuracy, spelling, and grammar by speeding through it.

- Complete each profile to 100 percent. According to Christine Cruzvergara of Handshake, completed profiles on the site are 80 percent more likely to receive a message from an employer. Completed profiles on all sites offer more information to employers, and they also demonstrate that you are someone who completes what you start.

- Consider all types of employment opportunities you come across: full-time, part-time, contract positions, fully remote jobs, gig employment, and more. There are pros and cons for each of these situations, but to overlook any category would be a mistake in today's job environment. Even if you are ideally seeking a full-time position, other situations might offer a way to get your foot in the door of an organization that really appeals to you.

- Search for jobs by a wide variety of key words. For example, if you're looking for roles in human resources, also search for "HR," "human capital," "talent," "people," and other terms you begin to notice in listings you are attracted to.

- Review job descriptions for key words to add to your résumé. As you know, many large employers use those Applicant Tracking Systems to search for key words in applications. According to one study, 75 percent of résumés are never read by a human. Make sure the words on your résumé match the key words in each job posting you apply for so the ATS will find you. The top reason an ATS would reject you is that your résumé doesn't reflect the qualifications of the job you're applying for.

- Follow all directions provided. Not doing so is a huge pet peeve of recruiters and can get you rejected by an ATS. If the posting says no phone calls, don't call. If the posting says to include a cover letter, include a cover letter. If the posting says to upload a Word document, don't upload a PDF. Don't make it easy for your application to be rejected for something avoidable.

In addition to researching jobs on general and niche job boards, it's also important to apply on individual employer websites. When you come across a desirable employer, go to their website to find out if they have a job board, then apply through their process. "Neanderthal" of the five hundred résumé submissions writes that, although many corporate website applications took almost two hours to fill out, he received 25 percent of his interviews from them. "Also, think long-term," he writes. "Some of these sites may have 10 worthwhile positions over the next six months. Put up with [creating an account and filling out your information] once, and from then on, it's as easy as clicking a mouse" to apply for each newly posted role.

Do Your Research to Stand Out from the Crowd

A common question I receive from college students participating in the campus recruiting process sounds something like, "How can I stand out to an employer when most of the other applicants are just like me? We all go to the same school, we all take similar classes, we all participate in extracurriculars, and we all have good grades."

The reality is, you can't always stand out by your past credentials. This is not just the case for students but also for recalculators making a major career change or reentering the workforce after a break, because your résumé won't "check all the boxes" on an employer's list. This is the perfect moment to remember recalculator rule #3: control what you can. You can't always stand out based on your experience or credentials, but you can *always* stand out by the amount of research you are willing to do on potential employers. Landing a job is not about you being the perfect candidate or the organization being the perfect employer. It's about convincing the employer that you're a perfect match for each other. Research will help you find that synergy.

From your networking experiences and any upskilling or reskilling you've done, you should have some sense of the issues and trends in the industry or industries you want to join. Now it's time to go even deeper with your knowledge. Even if you're a full-time student or unemployed or working a stopgap job to pay the bills, there's no reason you can't be consuming the same news and information as people in your desired organization or industry. One of my rules of thumb is that if I hear about a podcast, article, newsletter, or publication from more than one person in my field, I check it out. And, nowadays, the vast majority of this content is easily accessible on the Internet. Your goal should be to create your own personal newsfeed of career-related content.

Exercise: Create a Career Recalculation Newsfeed

I recommend that you use Twitter—though LinkedIn also works—to set up a personalized newsfeed of employers, industries, influencers, and topics related to your career recalculation goals. Twitter's List feature allows you to create a feed that is separate from any other accounts you follow on the platform. If you've never used Twitter before, you can quickly set up an account using the app or desktop version. You don't even have to use your real name or post a single tweet if you don't want to. This exercise is just about the information you will consume.

Begin by following every potential employer that interests you even a little bit. Then follow any professionals or leaders at those employers or in your desired field. Then follow any professional associations or industry publications related to the field. Once you've followed everyone and everything you can think of, you can do what I call "going down the rabbit hole" and scan through the lists of accounts that each organization or person in your newsfeed is following. This will alert you to new organizations and people you didn't even know about. Follow the ones that appeal to you. I encourage you to stretch yourself and follow at least fifty to one hundred accounts in total.

Once you've set up your feed, spend five to ten minutes each morning scrolling through it, just as you might start your day by scanning through the general news headlines. You don't have to read every word of every post, but you can look for trends and hot topics. Are there current industry issues being debated? What jargon and acronyms are used that you should be familiar with? Have any important leaders been hired or fired? What jobs are people and organizations posting about? (Many employers have separate Twitter accounts that exist only to post available jobs—you should also follow those if you're not already.) Think of yourself as a fly on the wall in all of these conversations. The best part is that this is all totally free and available 24/7. Companies spend millions of dollars a year on their social media presence; they want you to consume what they post.

When most people think of Twitter, they picture a feed of political tweets, celebrity selfies, or the opinions of friends and colleagues. But Twitter's own cofounder Ev Williams primarily thinks of the platform in the way I describe it: as a news system. "The way we started talking about it was as a real-time information network," he has said. We live in the Information Age, so you need to have as much information as possible to compete for opportunities.

Once you secure a job interview with a particular employer, take two more research actions immediately:

1. Spend at least an hour each day before your interview looking through the employer's website. Visit the press page to read current news, visit the leadership page to familiarize yourself with the names of the top executives, and generally become well versed in the products and services that this employer offers. If they have a company mission statement, commit its key elements to memory.

2. Set up a Google alert (google.com/alerts) with the employer's name. This will ensure that you'll be up-to-date with any media stories about the employer that might arise. Trust me, you don't want to miss a big announcement that takes place the day before your interview!

While you're doing your research daily, here are some ways to use the information you're consuming to support all aspects of your recalculation:

Find Reasons to Reach Out to Old and New Connections

As discussed in the networking chapter, it's rare that a job opportunity comes up in the first interaction with someone. The magic happens in the follow-up. But many people are unsure of what to say in subsequent communications other than, "Just following up!" This is where your research comes in. When you are keeping up with the news of potential employers and industries, you're bound to come across articles, opinions, and resources that you can share with the people already in your network. "I saw this article and I wanted to share it with you because it reminded me of our conversation" is a great way to reengage with someone. It's generous, it's relevant, and it demonstrates that you are in the know about important professional topics.

Discover New Potential Employers

You can't possibly know all the employers out there, especially when it comes to small and midsize organizations. Research can help you find hidden opportunities that other people might miss. This is how I landed my job at WorkingWoman .com. I was interested in jobs related to women and the Internet, and my former internship boss faxed (!) me a magazine article about the two major women's websites at that time, iVillage and Women.com. At the very end of that article, there was a sentence about some publishing executives from *Working Woman* magazine launching a potential third competitor. I cold-called the magazine's offices to ask if they were hiring for the new digital venture, and the rest is history.

Build Confidence and Depth of Knowledge

Imagine the feeling of walking into a classroom on the day of a big exam and you are totally prepared. You attended all the classes, did the homework, and studied the night before. When you start taking the exam, you are calm and confident about your answers. You don't have to fake anything or "wing" it. That's the feeling you want to have when you walk into a networking event or job interview. The best way to have that feeling is to keep doing your research.

If you actively follow the social media accounts of a potential employer and have reviewed their full website, you will walk into a job interview completely prepared. You'll speak fluidly about the various products or services of the company. You'll speak with clarity and specificity about why you're applying to work there and what you can contribute to their mission and goals. You'll have insightful, relevant questions for the interviewer. Overall, you'll come across not as an "outsider" job applicant but as an "insider" colleague-in-waiting.

Stephen Isherwood of the Institute of Student Employers in the U.K. used to be a campus recruiter for EY, and he regularly asked job candidates during interviews to talk about EY's history. "I wasn't testing them," he explains, "but I wanted to know how interested the candidate had been in learning about the firm. I want to hire somebody who is really interested in being here, especially when the work gets difficult."

Showing genuine curiosity through research is even more important to smaller companies, who usually don't have large employee onboarding operations or budgets. They also want to know that you cared enough to learn about them. One small business owner I know begins every job interview by asking the candidate, "Have you had a chance to look through our website?" If the answer is no, she ends the interview on the spot. There is no excuse not to do your homework and impress employers with your enthusiasm for their organization if you want them to bring you onboard.

The First Pancake: Embrace Imperfect Opportunities

Let's take a step back for a moment and think about what positions and employers you're even considering. I encourage you to be as open-minded as possible in deciding what jobs to apply for. Part of developing the skill of recalculating includes embracing the fact that, even if a job is not THE ONE, you can adapt and adjust to make it a great opportunity for your career. There is no perfect job or job search.

Take it from a card-carrying member of the Perfectionist Club: waiting for a flawless opportunity to magically arise will be your downfall if you don't let it go. You need to explore and embrace imperfect paths and cast the widest net possible to find opportunities, especially in a tight job market.

Dr. Lisa Stern, director of program impact and assessment for FourBlock, the nonprofit that provides career transition support for veterans, refers to someone's first job out of the military as their "first pancake," a metaphor shared with her by a veteran during Lisa's dissertation research. Maybe you're a better cook than I am (there's a 99 percent chance you are, given my skills in the kitchen), but the first pancake of any batch is usually a dud—it's overcooked, undercooked, awkwardly shaped, has too few or too many chocolate chips, or somehow gets flipped onto the floor instead of the plate. It's far from perfect, but you can still eat it (unless it's the floor pancake!) and be satisfied.

All of this is often true of your first "real" job out of college, your first position in a new industry after a career change, or, as with Lisa's clients, your first civilian role after leaving the military. It might be a dud. And that's okay. It will still be valuable.

Michael Abrams, the CEO of FourBlock, shared the story of a Marine Corps veteran, Rob, who completed his undergraduate degree in finance at a highly competitive institution and got a prestigious internship at a Wall Street bank. The following fall, Rob called Michael, very upset. "I really need to talk to you," he said. "I don't want to work on Wall Street. It's not for me. I need more balance."

"Congratulations, that's awesome!" said Michael. "Better that you realize that now in an internship than two or three years down the road in a full-time job!"

I love Michael's positive attitude about an internship or first job not working out. I'd even say your first few jobs after a big transition might not live up to expectations, and that's okay, too. Think back to our discussion in Chapter One about managing your expectations: if you go in expecting a new role to be the BEST JOB EVER, then you're sure to be disappointed. But if you go in hoping for your first job or jobs following a major transition to be stepping-stones and learning experiences, then you'll end up much happier and more fulfilled.

Here are some specific suggestions for widening the scope of opportunities to apply for by embracing imperfection:

Be Open to a Wider Variety of Industries

We are living in a time when even the largest of companies and industries rise, fall, and become disrupted faster than ever before. The average life span of a company on the S&P 500 (the five hundred largest publicly traded companies in the United States) is now less than twenty years, compared to almost sixty years in the 1950s. "Disruption is nothing new," read the report releasing these findings, but "the speed, complexity, and global nature of it is."

And this was *before* COVID-19.

Many industries are on the decline or were heavily impacted by the pandemic, including travel agencies, department stores, textile mills, print media, and cable TV. If you work in an industry like this, you'll likely have to consider a new field.

In addition to the extra profile elements LinkedIn offered at the beginning of COVID-19, which you read about in Chapter Three, in 2020 LinkedIn launched a new ongoing feature, #HiringNow, which is a frequently updated page on the network containing a running list of companies with open positions. Handshake did the same for entry-level employers. These types of lists are great resources that you can use to learn which companies and industries are thriving and hiring, even in challenging times.

When the pandemic first began, a majority of the employers on LinkedIn's #HiringNow list were, not surprisingly, large grocery chains. Many job seekers at the time told me they didn't want to work at a grocery store because they imagined themselves scanning items at the checkout counter. Sometimes that's the job, but I reminded them that those same grocery chains also had available jobs in marketing, human resources, e-commerce, analytics, finance, operations, and more.

Now is the time to let go of any stereotypes or potential misconceptions you have about various industries. To give another example, some people, especially millennials and Gen Zs, believe that industries like insurance or manufacturing are "old-fashioned," even though many employers in these fields offer great career paths and many tech-related jobs. I once spoke for an association of plumbing manufacturers and learned that they are eager to recruit workers who are passionate about environmental issues, particularly related to water conservation. That hadn't occurred to me! Bottom line: don't judge or disregard any industry without researching what it's really like to work in that field.

Consider a Wider Variety of Job Functions

Karen Ivy of Texas A&M–San Antonio shared a story about Sabrina, a second-semester junior who needed a job *fast*. She applied for anything that was available and was offered a job at a call center for GM Financial, which has a beautiful facility in San Antonio. "Some students think, ugh, I don't want to be on the phone all day," Karen told me. "But GMF offers part-time positions, they are very flexible, and they hire students. And it pays well." Sabrina took the job because she needed the paycheck, and while she was grateful to have it, she was not excited about "just answering phones."

It turned out that the job came with a lot of training, which Sabrina embraced. GMF encourages part-time employees to use the call center job and its accompanying training as a foundation for their careers. Sabrina took full advantage of the training opportunities in particular to bolster her communication skills. This put her on the radar of her manager, who promoted her to team leader, based on her performance and engagement in the optional trainings. Sabrina's manager also told the HR department that Sabrina was a good performer and quick learner.

As a result, an HR rep set up a meeting with Sabrina, who was brave enough to share that her ultimate career goal was to work in HR. She asked the rep questions about what skills were needed to get a job in that area, and she volunteered to help the department by giving tours to potential new hires. When a university relations position opened up in the human resources department, Sabrina applied and landed the full-time position.

"Sabrina's story is proof that companies want to retain great talent," Karen tells other students. "Even if you start with part-time work or a position you're not excited about, you can impress people and move internally." It's often better to get started in an imperfect job and build your reputation than to wait and wait and wait for the exact job you want. (We'll explore this even more in Chapter Six.)

Consider Other Trade-Offs

Embracing imperfection in your job search comes down to trade-offs. Think back to recalculator rule #4: know your nonnegotiables. Perhaps, like Sabrina, making money is the most important factor in your job search, so you're willing to consider a role that doesn't initially seem ideal. Or maybe you can't take a job unless it allows you the flexibility you need as a primary caretaker of children or elderly relatives. Or maybe feeling passionate about your work is an absolute must, even if it means taking a pay cut. (One study found that 64 percent of millennials would rather make $40,000 at a job they love than $100,000 at a job they find boring.) All of these are legitimate requirements and I will never ask you to bend on your "musts." But I will ask you to consider compromising in other areas.

Stacey Delo, coauthor of *Your Turn: Careers, Kids, and Comebacks—A Working Mother's Guide*, and leader of a job platform and community for women returning to work after time off, prefers the term "priorities" to "trade-offs." "Your priorities will change and don't have to be permanent," she says. "If your work priority right now is flexibility because it allows you to manage a different part of your life, look at that as a short-term prospect. In a perfect world, you would take a flexible job that's high paying, but maybe it's okay to earn less for a bit now because you're gaining something else and the flexibility will pay off in the long run." If you are going to prioritize one element of a job over salary, Stacey recommends that you do set a minimum threshold of pay in order to establish boundaries.

Exercise: The Art of the Trade-Off

Take some time to prioritize various elements of your ideal job and their overall importance to you. This will help you to make potential trade-offs when offers start coming in. Fill out each section of the list below, then rank each factor in order of importance from 1 to 8 (or 9 if you add an "other"), so that 1 is the most important priority in your next job and 8 or 9 is the least important factor and possibly one that you would consider compromising on.

____ **Job Function**

The job function of my next position must include (examples: work I'm passionate about, challenging work, helping people):

_____

_____

_____

____ **Employer/Culture**

The employer of my next job must be (examples: a brand-name company, environmentally friendly, committed to diversity, equity, and inclusion): _____

_____

_____

_____

____ **Salary**

The salary of my next job must be no less than $_____

___ Benefits

The benefits offered by my next job must include (examples: health insurance, a 401(k) match, vision and dental, X amount of PTO): _____

_____

_____

___ Title

The title of my next job must be at least: _____

_____

_____

___ Location

The location of my job must be (examples: no more than thirty miles from my home, remote, in a city, near public transportation):

_____

_____

___ Schedule

The schedule of my next job must be (examples: flexible, 9:00 A.M. to 5:00 P.M., no weekends): _____

_____

_____

___ Other: _____

My next job must include:

_____

_____

_____

Conquer Career Fairs

Talk about economies of scale—a career fair allows you to potentially meet dozens of employers in one place, whether in person or virtually. But you're not the only one jumping at the chance, which means you have to do some preplanning to make the most of your time when you attend. To find job fairs in your geographic area or related to your industry or career level, check out the directory at jobfairsin.com.

An in-person job fair usually consists of a large event where each employer has a booth or table. You walk around the event, wait in line, then have a few minutes to make a connection and drop off your résumé. Virtual career fairs generally involve one of two options: you can either register for a virtual information session with an employer, which is similar to attending a webinar that includes a Q&A via the chat feature or attendees unmuting themselves to ask questions; or you can register for a one-on-one video meeting with an employer during the designated hours of the fair. The group setting is best for when you are just exploring a particular employer or industry, and the one-on-one format should be your choice when you are very interested in working for that organization.

Here are some tips for navigating the various types of job fair situations:

- **Make a plan:** For in-person fairs, visit the website of the sponsoring organization to look for an event map so you can design a visitation strategy. You want to meet your preferred choices early on when you—and the recruiters—likely have the most energy. For virtual fairs, register as early as possible for the type of sessions and time slots you want. A virtual career fair can be more valuable in some cases because you'll have a designated amount of one-on-one time for each recruiter conversation you sign up for rather than jostling with other candidates to keep a recruiter's attention with a long line of people waiting behind you.

- **Do your research:** Know which companies will be at any career fair and research each one on social media and their own websites, so you can know as much as possible about each organization. Don't overlook the employers you've never heard of—remember to keep an open mind and consider all types of organizations and industries. An advantage of virtual career fairs is that you can keep your notes in front of you while speaking with a particular recruiter. One great tip is to put a few pointers on a sticky note and attach it to your monitor out of sight. It will prevent you from looking down to consult your notes and keep your gaze steady.

- **Create customized résumés:** If you only have a few target employers, go the extra mile and create customized résumés for each one—and potentially even personalized cover letters—to submit to your primary targets. When I was job hunting, I would organize my résumés into different colored folders—one for each employer on my target list—to keep everything organized.

- **Request contact information:** You don't want to rely on the recruiter seeking you out after the event. Make sure to collect a business card or write down the name and contact information of the company representative you've met (it is totally appropriate to ask for this information) so you can follow up with a thank-you email. It's also a good idea to ask recruiters if it's okay to connect with them on LinkedIn. Some will say no, but the recruiters who say yes will know to be on the lookout for your connection request.

- **Prepare a short "elevator" speech:** This is essential for one-on-one virtual career fair appointments. For IRL fairs, note that different events have different formats; often you might be just dropping off your résumé. But if you are able to access a few minutes of a recruiter's undivided attention, be ready with a short introduction, based on your personal brand and career story from Chapter Three, that explains who you are and why you're particularly interested in that company.

- **Prepare questions:** "You can make such a positive impression by asking smart questions," says Christine Cruzvergara of Handshake. "By doing your research, you can ask deeper, more thoughtful questions: Where is the business going? What is your opinion on industry trends? How does this particular role help to address challenges the business is facing? Asking questions is highly underrated." Remember my grandfather's advice about listening twice as much as you talk.

- **Jot down some notes:** Your chats with various employers might start to blend together, so after each conversation, write down the recruiters' names and highlights of each interaction in your phone or a notebook so you can keep track and refer to these notes in subsequent outreach. Taking notes also provides a visual sign to recruiters that you are an engaged and enthusiastic job candidate.

- **Continue the conversation over social media:** Reach out to select recruiters on LinkedIn (if they've agreed to connect) or follow their personal and/or company Twitter accounts. Smaller employers in particular will be impressed at your initiative, and you will have an inside track to professional and company news that can help you pursue future opportunities.

Remember that a career fair is just one facet of a comprehensive job search, but it's an excellent opportunity to make a lot of contacts and hear about potential openings all at one time. Solid preparation and follow-up will help you make the most of the occasion.

Make Recruiters and Hiring Managers Your Allies

As you are narrowing in on organizations you want to work for and specific jobs you want, it's important to build relationships with the recruiters who are the gatekeepers for those jobs. When you apply to many midsize and most large companies, the recruiter will be the person who positions you to the hiring manager (a.k.a. your potential future boss).

Alyssa Welch, an HR practitioner who has held recruiting positions in multiple organizations, says that getting to know recruiters is a critical differentiator, especially for job seekers whose backgrounds don't align with the positions they want.

According to Alyssa, most job candidates treat recruiters as logistical coordinators, only reaching out to schedule an interview or to ask when they will find out if they got the position.

Instead, she recommends that you connect with recruiters to ask about the culture of their organizations and to build their confidence in recommending you to a hiring manager. "By spending time with that recruiter you go beyond what's prescribed," she says. "You're no longer just another résumé. They can say they talked to you and mention something that stands out about you. Being able to show up with a star candidate makes the recruiter look good—they will be selling you! I always felt better proposing a candidate to a hiring manager when I felt I knew the candidate a bit."

The additional benefit to making a positive impression on a recruiter is that they will keep you in mind if one particular job doesn't work out and a similar future role opens up.

For example, Lauren was a marketing executive in healthcare wanting to transition to a similar role in the technology industry. She had always been interested in working for a specific big tech company, so she set up alerts on LinkedIn for any marketing jobs posted by that organization. One day in January 2020, she saw a position that piqued her interest and aligned with her marketing leadership experience.

Lauren had never worked in the tech industry and knew she'd be screened out immediately if she applied online, even though she believed her marketing skills and experience were transferable. Instead of risking being disqualified for a lack of industry experience, she searched through the company's employees on LinkedIn until she zeroed in on the person she was almost certain would be the hiring manager for this position: Yoshi. She decided to send Yoshi a direct message. Pro tip: Because she didn't have the LinkedIn premium account and couldn't reach him without it, she signed up for the free trial just to send that one message.

"I thought, 'If it works, great. If it doesn't, what's the downside?'" Lauren says.

Yoshi replied within twenty-four hours and said he wasn't the right contact but directed Lauren to the correct person to reach out to: Monica.

Lauren immediately reached out to Monica. Although she didn't hear back from Monica for a full month, Lauren followed up and got a response the next day. (Always follow up!) It turned out the delay was due to the fact that Monica's team had been going through a big reorganization—as is so common these days—and Monica hadn't been sure which positions would still be available after the fact. "If you're interested in other similar opportunities, we'd love to talk to you," she wrote in her reply.

Monica connected Lauren to an internal recruiter who then referred her to another open position. While it wasn't a perfect fit, Lauren really wanted to get a foot in the door at her dream tech company, so she applied. By this point it was March 2020; the pandemic had hit and Lauren's healthcare employer was starting to furlough employees. She was asked to furlough several members of her team. This seemed like a tough moment to abandon ship, but the tech company was on a hiring spree and Lauren felt that this was her best chance to work there.

Although Lauren was offered the not-perfect open position

and planned to take it, the recruiter then invited her to interview for her original dream role, which was now open and available. She interviewed, got the job, and began working—I love this part—for Monica, the exact woman she had been referred to by her first contact at the company, Yoshi. Recalculation is rarely a straight line.

This success story began with the risk Lauren took in contacting the person she thought *might* be the hiring manager for her dream job. What did she say in the fateful message that helped her change industries and land her dream position in the midst of a global pandemic?

"I directly called out my lack of a tech background," Lauren explains. "I told them I've never sent a message like this one but I was compelled by the job and the company. I said that on paper I might not look like the ideal candidate, but here's why I think I would be an excellent candidate to join your team. Here is the value I can bring to your customers."

Lauren's story is such a great reminder that you only have to impress one person to get a foot in the door of any opportunity.

The New Rules of Job Interviews

When the COVID-19 lockdowns were first imposed, employers with open positions immediately moved almost all job interviews to a virtual format. Even the U.S. Army held its first-ever virtual hiring campaign to recruit ten thousand new soldiers. Given the cost savings and relative ease of the transition, many companies have made this change permanent. While some employers still prefer to conduct in-person final-round interviews when possible, numerous recruiters have told me that the move to virtual interviewing was inevitable; the pandemic just pushed this change to happen faster.

If you're a millennial or Gen Z job seeker, you might actually prefer virtual interviews, thanks to being raised in the era of smart technology and the Internet. If you're older or less technologically savvy, this format can take some getting used to. But anyone can learn to be authentic and impressive on camera with a little preparation. Here are some best practices for in-person, phone, virtual, and recorded video interviews, as you'll likely face some combination of all of these formats during your job search:

Turn On Your Camera for Live Virtual Interviews.

I didn't think I'd have to write this, but several recruiters told me they frequently have candidates ask if they need to use the camera. Yes, you do! Eye contact and confident body language like good posture are important parts of making a positive impression, even through a computer screen.

Ask About Appropriate Dress.

Before any live, recorded, or virtual interview, ask your main contact, usually the recruiter, about appropriate attire. There are just no rules or common standards around dress anymore. When I started my career in the mid-'90s, it was expected that everyone should wear a suit (and pantyhose for women, OMG!) for almost any professional job interview. That has changed significantly over the years, and changed again when interviews became primarily virtual. Some organizations that once expected interviewees to wear suits now feel that it's silly to do that when you're at home. Others still want the suit. The only way to find out is to ask.

No matter what the designated attire for a virtual interview, I recommend dressing fully, head to toe, as if you were attending the interview in person. It's fun to wear a suit on top and sweatpants, shorts, or yoga pants on the bottom, but you'll feel more professional and confident if you're fully dressed and wearing shoes. And, if for any reason you need to stand up during the interview, you won't be caught off guard.

For Virtual Interviews, Check Your Tech.

Michelle Horton of Wake Forest University School of Business advises job seekers to always ask employers which technology they are using for their interviews. You can ask this early in the process so you have time to practice with any platform before it's time to use it. "Take away the anxiety by getting as much information and practice as possible," she says.

While some organizations have their own proprietary interviewing software, most of the common platforms, such as Zoom and Hirevue, provide the opportunity for you to practice through free demos. Be sure to do this before any formal interview so there are no surprises or issues on the big day. Likewise, make time before any virtual interview to double-check that your Internet connection is strong, your device is charged (whether it's a laptop, tablet, or phone), and you have all other applications closed so you're not distracted by pop-ups or "dings" during the interview.

Craft Your Virtual Environment.

For recorded and live virtual interviews, you must curate your environment and self-presentation to make an impression that is consistent with the personal brand you want to put forth. Remember recalculator rule #3: control what you can. According to Ian Siegel, cofounder and CEO of ZipRecruiter, "You get one second, the moment you first come into frame, when they're going to develop a whole idea of who you are, and it's incredibly difficult to get them off that. You have full control over that one second." Your first moment on-screen is as important as anything you say afterward.

Michelle Horton of Wake Forest tells her students and alums that how they show up virtually should be as close as possible to how they would show up in person. "The rules are the same," she says. "Don't get lax!" Would you have a smile on your face and make eye contact when you first meet an interviewer? Yes, so make sure you are smiling and looking directly at the camera when the interviewer first sees you on-screen. Would a roommate or family member be in the interview room with you or making noise in the next room? No, so do your best to be alone for a virtual interview. (If you can't control who else is there, I would politely alert the interviewer at the beginning of your session that you are not able to be alone for the interview and there might be some background noise. They will understand that sometimes your environment is not fully in your control.)

You don't want anything to distract the interviewer from paying attention to you and your brilliance, so make sure your environment is as simple and professional as possible. Sit on a plain chair (not on a couch or perched on your bed) and have a simple background like a plain wall with a bookcase or some art. Sit facing a window or buy an inexpensive ring light to make sure your face is well lit. Check your camera angle in advance to make sure you're well framed with your head and torso showing and you're not looking down at the screen. I like to stack my laptop on books to get a better angle.

If your background is too busy and you can't change it, then find out if the interview software offers simple virtual backgrounds, such as a neutral color or an office environment background—this is a great option on Zoom.

Be Mindful of Your Body Language.

Both interviewing in person and being on camera can feel unnatural, so practice in advance until you feel more comfortable. Habits like fidgeting, twirling your hair, using upspeak (when every sentence sounds like a question), banging on the desk to make a point, or using verbal fillers (such as "um" or "like") can be unprofessional and distracting during an interview—and they're often more noticeable when you're on a screen. Sometimes you don't even know you have a particular habit until someone points it out or you observe it on video. (For me, it was tilting my head to one side. Once I noticed, it was easy to correct.)

The best way to discover if you have any of these habits is to do a mock interview, whether in person with a family member or friend, virtually with a mentor or university career services professional, or alone by recording your responses on your phone. Speaking of the phone, body language matters on telephone interviews as well. Even when an interviewer can't see you, they can hear the energy (or lack of it) in your voice. I recommend standing up for phone interviews and positioning yourself in front of a mirror so you can make sure you're smiling and look positive and attentive. (I do this when I'm presenting webinars, and it makes a big difference.)

Prepare Stories to Answer Behavioral Interview Questions.

If you've had a job interview in recent years, you've likely been asked behavioral interview questions—those that begin with "Tell me about a time when . . ." These questions can also take the form of STAR: describe the situation you were in, the task you were asked to accomplish, the actions you took, and the results you achieved. The goal is to show, not tell, what you have accomplished with specific examples. As a career changer, recent grad, or work returner, be mindful to explain how experiences in one realm—in a different industry or as a student or a stay-at-home parent—translate to the industry and job you are applying for. Don't rely on the interviewer to make that connection.

To demonstrate this approach, let's return to Lauren's story. To prepare for her virtual interviews with the large tech company—which included six hours of virtual interviews in a single day!—she wrote out twenty-five stories of her accomplishments from her healthcare career (in the form of STAR) and how they could be relevant for the tech sector. She reviewed these stories every day for three days before her interviews so that she would be able to recall them conversationally, depending on what questions each interviewer asked.

On the day of the interviews, she put the stories up on her computer screen and deleted each one as she told it, so as not to repeat any examples. My hunch is that her preparation would have served her just as well if she'd done in-person interviews without having these stories right in front of her, but she took advantage of the virtual setup to keep her notes handy. Clearly, she nailed the interviews.

Take Notes and Ask Questions.

Another suggestion from Ian Siegel of ZipRecruiter is to visibly take notes during a virtual job interview. The same holds true for an in-person interview. As we've discussed, taking notes is a visual cue to the other person that you are paying attention and eager to learn. Siegel takes this a step further, recommending that, when the interviewer asks if you have any questions, you refer to your notes and ask a question based on something the interviewer said during your conversation to demonstrate how engaged you've been.

You should also prepare questions based on the Internet research and human networking you've done during your job search. Rather than posing generic questions such as "What's a typical day like?" or "How did you decide to work for XYZ Company?" use your questions to demonstrate the knowledge you've been building. Ask questions about the direction of the business, specific initiatives you've read about, or timely topics in your industry. The goal is to sound like a colleague who can hit the ground running on day one if you are hired.

Another tactic is to pose questions that will help you best position yourself during the next steps of the job search process. Alexandra Carter, director of the Mediation Clinic at Columbia Law School and author of *Ask for More: 10 Questions to Negotiate Anything*, recommends asking questions such as "Tell me your biggest need for this position," or "Tell me about the last superstar hired and what made them so great." Questions like these will help you target the rest of your interview answers and, potentially, determine what you should emphasize about yourself when negotiating for salary, benefits, and more. Remember that recalculation can take place during the various stages of the job interview process itself.

Don't Curb Your Enthusiasm.

With all respect to Larry David, enthusiasm is a critical component of a successful job interview, in any format. Don't be afraid to say, particularly at the beginning and end of any interview, "I am so excited to be considered for this role" and to explain exactly why. Enthusiasm is especially important for recalculators, because you may not be a "traditional" candidate for a particular role. Genuine passion and excitement are important qualities if you are a recent college graduate, who may be lacking work experience; an older job candidate, who may fear an interviewer will have a bias against your age and energy levels; or a mid-career returner, who is rejoining the full-time working world after time off. Why is enthusiasm so important to employers? In the words of golfer Lee Trevino, "You can't teach passion. You can teach everything else."

Be Calm If Anything Goes Wrong.

One of Michelle Horton's colleagues participated in a Zoom job interview, and when she tried to share her screen for a slideshow presentation, it didn't work. Instead of panicking, she went into problem-solving mode, resolved the issue, and continued with her slideshow. The interviewers saw in real time how she handled a problem by remaining calm (even if she was freaking out on the inside).

Employers know that interviewing is stressful, both in person and virtually, so they are not just listening to your answers but also noticing how you handle the stress of the situation. (I think we can all agree that dealing with challenges is a big part of working—and living—in today's world.) If something goes wrong during an interview, take a deep breath and don't panic. Being able to think on your feet is a valuable skill that will impress your interviewers and better your chances of getting the job.

Consider the Bigger Picture.

One final thought on interviews: remember that they are also a form of networking. Even if you don't land the particular job you're interviewing for in this moment, you are making an impression on a new person or group of people every time you interview. Jessica Scott, the counselor who returned to work after six years away, embraces this view. "Every time you interview, it's a chance to get in front of people and talk about yourself and the value that you bring," she says. In fact, she parlayed a job interview for a position she didn't get into a three-month position covering for someone on maternity leave many months later. The interviewer remembered her and, although she wasn't the right fit for the job she had interviewed for, Jessica had made enough of a positive impression to be kept in mind for future opportunities.

Exercise: Practice Makes Progress

The key to job interview success is overpreparation. While you can't predict every single question an interviewer will raise, you can anticipate some of the frequently asked ones and be fully prepared to impress with your replies. Here are three of the most common questions and recommendations for answering them. Put your own spin on your answers using these guidelines and then practice, practice, practice:

1. Tell me about yourself and why you want this position.
This deceptively simple question is probably the most important one. You need to concisely make your case for why you and the job in question are a great match. Alyssa Welch, the HR consultant, recommends answering this question in three parts:

- Here's what I know how to do. (Briefly describe your background and skills and how they led you to this moment, a.k.a. your career story.)
- Here's what I'm interested in. (Demonstrate your passion for the type of work this job requires and show off that you've done your homework on the interviewer, role, company, and industry.)
- Here's how I can meet your needs. (Explain how you can achieve the goals of the position—this is again where your research comes in. Demonstrate that you've done your homework through informational interviews, social media channels, and other sources to truly understand how this role fits with the organization's overall mission.)

Remember to keep your answer short—just a couple of minutes—to indicate that you can communicate your story concisely yet effectively.

Finally, if your career story involves a gap or transition in education or employment due to COVID-19, you should be prepared to talk about what you did during that period of time. Did you read more books? Volunteer for a cause you care about? Take care of a family member? I don't believe there are any wrong answers, but I would recommend anticipating that a recruiter might ask about your pandemic experience.

2. What are your biggest strengths?

In your mind, you should interpret this question as "What are your biggest strengths *that match the strengths we asked for in the job description*?" Never forget that you are not making the case for yourself generically; you are making the case for yourself for this specific position. In challenging times like we find ourselves in, you can also use your reply to this question to discuss your adaptability to volatile circumstances.

Be sure to give a specific example of each strength you mention, such as: "One of my biggest strengths is flexibility. When the pandemic first started to spread and I was furloughed from my previous employer, I called all of my favorite local businesses and asked if I could help them with any marketing projects. I ended up assisting businesses in three different industries—a café, a dry cleaner, and an auto-body shop. I'm excited to apply that learning to this marketing manager position."

3. What are your biggest weaknesses?

Lots of people think the best way to answer this universally disliked question is to take a strength and pretend it's a weakness: "I am just so punctual. People always complain that I'm early!" or "I'm too much of a workaholic and never quit until a job is complete!" Eye roll. People can see right through those kinds of answers. The better approach is to use this question to demonstrate your growth mindset. Show that you are self-aware enough to know that you are not the best at everything and that you are already taking steps to improve as needed. For example: "One skill I've been working on improving is public speaking. During my time out of the workforce, I joined an online Toastmasters club and signed up to teach a workshop at my church in order to get better at presenting in front of large groups." Note that this answer does not include, "I'm terrible at public speaking!" Focus on the positive.

I encourage you to search online for additional interview questions to practice answering, particularly questions that are specific to your industry or desired job function. Some larger companies will even provide candidates with sample questions or case studies to help you prepare. Take advantage of it all.

Follow Up, Follow Up, Follow Up

Once the interview (or multiple interviews, as is the case with many employers these days) comes to an end, you can take a breath, but don't rest for too long. Follow-up is critical. First and foremost is a thank-you email.

I usually try to be diplomatic when people make mistakes in the job search process because—as I mentioned at the start of this chapter—it is wildly unfair how little most people are taught about how to land a job. But I am constantly baffled by the fact that most job seekers don't send a thank-you note after interviewing for a job. This has been a hard-and-fast rule of job hunting since forever. You must, must, *must* do this within twelve to twenty-four hours after interviewing for a job. Ten or fifteen years ago I would recommend mailing out a handwritten note with a real live stamp, but that's not realistic anymore with so many people working remotely or traveling frequently. An email is the way to go, and the sooner the better. You want to appear eager and excited about the role.

The key to a great thank-you email is remembering that this is another reflection of your personal brand and another (possibly final) opportunity to make a fantastic impression and reiterate why you're the best person for the job. Check that you are spelling the person's name correctly, say thank you, mention an interesting or helpful detail that was brought up during the interview and relates to your being a great fit for the job, and reiterate your enthusiasm. Then sign off. Your note doesn't have to be long; it just has to be timely and appropriate.

If you met with three people, you should send three different thank-you emails. Don't just copy and paste the same message. Make sure each note contains a personal nugget, something that only you could have written and only to that person. This demonstrates that you pay attention to details and will be a thoughtful colleague. (I know of one candidate who lost an opportunity because he sent the exact same thank-you note, word for word except for their names, to the five people he interviewed with. They were not impressed.)

After that, be patient. Wait at least seven to ten days before making any additional contact with a company once you've sent your post-interview thank-you email(s). It often takes at least this long, and sometimes longer, for a decision to be made, particularly at a large company and particularly in challenging economic times when a lot is going on.

If you don't hear from an employer after ten days or in the time frame they specified (which you are welcome to inquire about at the conclusion of your interview), try sending an email or picking up the phone. Robin Solow, the HR professional and recalculator, reminds job candidates to think about what is happening in the recruiter's world. At times, she told me, she would be recruiting for up to fifty jobs—that's a lot of people to keep track of. "Sometimes candidates just fall through the cracks," she admits. "Oftentimes it's helpful to follow up because it will jog the recruiter's memory. Also keep in mind that the strategy of the business might change. Jobs go on hold all the time." While recruiters try to be communicative and transparent, sometimes it's just not possible.

What Robin recommends, from her experience as both a recruiter and a job seeker, is to try different methods of following up—voice mail, email, and even texting if the recruiter has texted you first. "Make it fresh every time," she advises. Instead of writing or saying, "Following up again!" do some additional research on the company and point out another fact you've learned that makes you eager to contribute to the organization's success. Keep your tone positive, professional, and appreciative. "During my job search," Robin says, "a recruiter recommended that every time I send a communication, I should read it out loud and think about how the message is going to be received by the person before I hit send." Wise advice for all communications, not just those to recruiters, I would say!

The Million-Dollar Question: A Recalculator's Salary Negotiation FAQ

I have every confidence that the hard work you put into a job search will pay off in real job offers. The final step of the process is exciting, but—let's be honest—can also be dicey: the salary negotiation. "What are your salary expectations?" is some people's most dreaded question in the job search process. Here are answers to some common salary negotiation questions for recalculators:

Q: How do you determine what salary range is appropriate?

A: When you're newly graduated, transitioning into a new industry, or have been out of the workforce awhile, it's vital to conduct research to get a feel for pertinent salary ranges. Fortunately, there are sites galore that track the going rates for various positions in different locations, which can be an important factor since geography can make a big difference in rates. Check out the calculators and resources at sites like Glassdoor, PayScale, Salary.com, and SalaryExpert to give you a ballpark.

Of course, websites can provide only generic information, and there is a lot of nuance to salary numbers. Although it can feel awkward, I strongly encourage you to discuss salary ranges with people in your network as well. This includes university career services professionals, informational interviewees, and recruiters. You might phrase it something like this: "Based on your knowledge of the industry, what do you believe is an appropriate salary range for the kinds of positions I'm applying for?" If you ask a variety of people, you'll have an easier time averaging out a suitable request.

Remember, too, that an employer is most likely to offer you the lower end of any range you provide. When I started out, I made the wrong assumption that people would offer the average of any range I provided—nope!

Q: Do I have to throw out the first number?

A: In any negotiation, you never want to go first; unfortunately employers know that rule, too, so they'll try to get you to name your number. It's tricky, of course: underestimate and you might lock yourself in below what they could offer; go too high and they might rethink your candidacy. The best approach is to do extensive research and follow the advice from the first question to be very clear in your own mind on your salary needs. If the company can't even offer you the minimum pay you need, it's a no-go.

Q: What if they ask my current salary, but it's nowhere near where I want to be?

A: This can be a bummer of a question, because if you're stuck in a salary basement, this request can make you feel like you'll never rise above it. Fortunately, this question is illegal in a growing number of states and municipalities. But even if it's not illegal where you live, you can avoid a direct answer if you'd prefer not to divulge your earnings. You can dodge the question by saying your request for this position is in such-and-such a range and that your previous salary isn't as relevant because you're switching careers or locations or changing responsibilities, or whatever your situation may be.

On the flip side, your past salary might be higher than what you expect in this role because you are deliberately taking a step back. Some recalculators worry they will be discounted for being perceived as "overqualified" or too expensive. In this case, Alexandra Carter of the Mediation Clinic at Columbia Law School recommends saying something like this: "I'm aware that I'm relatively senior. I want you to know that at this point in my career I'm really looking for new challenges and the ability to work hard in the right place. That is what's most important to me."

Q: How do I avoid being lowballed, especially if I'm a career changer or nontraditional candidate?

A: If you're making a career transition, the hiring
manager might take that as a signal that you're willing
to do anything to get your foot in the door—even accept
a below-market salary. That's where you have to make
your case that even if you don't have direct experience,
your transferable skills will allow you to hit the ground
running and achieve the goals of the position. Alexandra
Carter recommends countering any hesitation or
resistance by asking, "What are your concerns?" Then
you can highlight specific arguments in favor of a higher
salary.

Remember you might have to play the long game. If
an employer does offer a low figure and won't budge,
treat that as just the start of your total compensation
package. You can ask for other benefits that might be
as valuable to you as money, such as a more senior title,
the ability to work remotely, more flexible hours, extra
vacation time, tuition reimbursement for classes you'd
like to take, or even assistance with student loans. Salary
is just one component of your total package; in fact,
benefits comprise an average of nearly 40 percent of total
compensation. Don't assume that anything is off the table
until you ask. You can also ask to schedule a check-in
in a few months when you've had time to prove yourself
and can revisit the salary issue. If they agree, have them
commit in writing to a salary review at that time.

Finally, you might eventually realize that this field is in
fact lower paying than you thought; or if you're returning

to work after a long break, it could be that the only company willing to take a chance on you is not exactly known for its generosity. The reality may be that you'll need to vow to work as hard as you can, learn as much as you can, and then parlay that newfound experience into a salary bump at a new company as soon as you can. We'll work on this in the next chapter.

I know that job hunting can be a stressful process with many ups and downs, but remember that each time you experience it, you are getting better and better at this essential skill. And, if and when you are on the other side of the hiring conversation in the future, you'll have a lot of empathy for the candidates who hope to work for you.

Turn Any Job into a Great Job

**Transform Every Position into an
Opportunity to Build Your Skills, Network,
Self-Awareness, and Future Opportunities**

If opportunity doesn't knock, build a door.
—MILTON BERLE

Congratulations!

You accepted an offer and started your dream job.

Or, you accepted an offer and started a job that seems mostly okay, if maybe not perfect.

Or, you're still working in the exact same job as when you started reading this book, but you've decided to recommit to your current role.

Congratulations on any of the above recalculations!

No matter how you got to your current job or how long you plan to stay, you can turn any position into a valuable stop on your career journey and a stepping-stone to your next opportunity. As always, your attitude and approach make all the difference. Reflecting back on the chapters leading up to this point, any job can improve your overall mindset, your clarity on the path you're forging, the career story you want to tell, your contacts and network, and your ability to succeed in future job searches. Your mission now is to approach your current job in a positive, proactive way to make future recalculations a little easier.

Here are some of the benefits you can get out of any job:

- Skills: Use what you learn to help build your résumé and career story.
- People: Meet new professional mentors and networking connections, and cultivate personal relationships. (Not for nothing, a lot of people meet their significant others at work!)
- Self-knowledge: Discover work that you didn't expect to enjoy or find fulfilling, and/or find out what work you never want to do again. Learning what you don't want can be just as important as figuring out what you do want.
- Promotions and internal mobility: Sometimes it's worth taking a job at a good company, even if the role doesn't perfectly fit your interests. This allows you to get a foot in the door, and once other jobs become available, you can potentially move into a more fulfilling or lucrative position.

This chapter will guide you through each of the above outcomes (okay, except meeting a significant other) in the sections below. Note, however, that the advice in this chapter to make the best of any job is not meant to gloss over a discriminatory work environment or an abusive boss. Those are serious issues that point to an organization's larger workplace culture, and they cannot be resolved through a change in mindset or more networking. If you are facing such a situation, please reach out to a trusted professional for guidance on how to report the situation through the proper channels—and, of course, leave the job if you need to for your mental or physical well-being.

Every Job Can Expand Your Skills

There are so many ways to learn, grow, and develop while working in any job. This includes the "soft" and "hard" skills we explored in Chapter Three.

Even if your current role is not ideal, think about ways you can turn it into a "curriculum" of growth that will help position you for your next opportunity. Refer to the growth mindset exercise in Chapter One and the areas you wanted to improve about yourself. If you want to upgrade your writing, save the well-written emails you receive and start borrowing some of the style that impresses you. If you want to be promoted to a management role in your next job, keep notes on the actions and characteristics you like and don't like about your current boss. If you want to get better at work/life integration, seek out mentors who seem to blend the two well.

Sometimes you'll have to directly ask for growth opportunities, and I encourage you to do so. Millennials who graduated into the global financial crisis of 2008 offered similar advice to Gen Zs who were graduating into the pandemic-induced recession of 2020. As one class of 2008 college grad shared, "Instead of asking for a fancy salary or a fancy office or something, ask for access to certain meetings or certain responsibilities—things that would act as résumé items two or three years down the line."

Many times, gaining new skills and unlocking advancement opportunities can be as simple as recommitting to the basics. There is a poster I've seen on the bulletin boards of countless university career centers that lists "10 Things That Require Zero Talent." It's a fabulous reminder of all the ways you can stand out and get noticed as a professional in any situation, in any job, and at any stage of your career without formal training or financial investment. I've also seen this referred to as "doing the boring stuff uncommonly well," but I would argue that these actions are far from boring—they are indispensable.

These ten things are particularly impactful ways to make a positive first impression in a new job, but they can also improve your standing if you are already established in your organization:

1. Being on Time

Do you struggle to be on time for appointments, meetings, or video calls? Use your current job as an opportunity to start improving on your punctuality. Richard Branson, the founder of Virgin Group, identifies this as his greatest productivity hack. "If you want to be more productive, then start at the start: get there on time," he has written. "Whether it is a meeting, a flight, an appointment, or a date, ensure you are there when you say you will be there. . . . Being on time is respectful to your hosts and also means you can effectively manage your day." We've all had that terrible feeling of joining a meeting where everyone's video is on and you are still scrambling. When you're on time, or ideally a few minutes early, you'll be more relaxed, more prepared, and more focused. You may even stumble upon some new opportunities, such as getting in some face time with the CEO a few minutes before everyone else arrives or logs on.

2. Making an Effort

You can improve your day-to-day happiness and fulfillment, even in a job that doesn't exactly thrill you, by making a concerted effort to do more than the bare minimum required to earn a paycheck. This might mean chatting more with customers rather than just silently ringing up their purchases, participating in that Zoom happy hour with colleagues instead of logging off right at 5:00 P.M., or learning more about the work your organization does rather than just processing invoices.

In college, I had a summer job as a counselor at a day camp, and I was disappointed to be assigned to a group of five-year-olds when I had hoped to be with older kids. For the first week, I just sat on a bench and supervised their play. I was doing my job, but I was bored and unhappy. Then I realized that I had the choice to sit and be bored, or try to make the best of it. So I jumped into the sandbox and had a whole lot more fun playing with the kids than just watching them from a distance. The days went by much faster, too. Except for one very unfortunate bee sting, it turned out to be a pretty good summer job.

3. Being High Energy

While not everyone has a high-energy personality, you can make it a goal to bring more enthusiasm to certain projects or situations or better manage your energy levels. We talked in Chapter Five about how important enthusiasm is during a job search; it can be equally important on the job, especially in challenging times. Higher-ups notice who walks sluggishly into the office compared to those who walk tall and greet their colleagues with a friendly smile. Clients witness who is slouching on a video call versus who is eagerly participating. Even a small improvement in your liveliness can make a big difference in how you are perceived, what projects you're asked to join, and whether you are offered opportunities for advancement.

No one expects you to be perky all day long, so start to measure when your energy is at its highest and lowest (I know I have way more energy in the morning and then it dips in the afternoon) and experiment with scheduling your most important tasks or interactions for your higher-energy hours.

4. Having a Positive Attitude

Another way to improve a job and open yourself to new opportunities is by improving your attitude. You can do this by changing how you approach mundane or unpleasant responsibilities or adjusting how you interact with colleagues and higher-ups.

In high school, I worked at a video store (retro, right?) and my manager eventually gave me her computer password so I could take care of more tasks. Her password? "SSDD," for "Same Sh*t, Different Day." I always thought it was funny, but thinking back on it now, I imagine her feelings about the job got worse and worse every time she typed in that password, which was probably dozens of times a day. If you put your attention on the negative, you'll see negativity all around you.

Do you have any similar negative habits, like displaying sarcastic work quotes on the bulletin board above your desk or starting every call with a colleague by complaining about your boss? I'm not saying you need to be Pollyanna, but having a more positive attitude, like improving your energy levels, is a quick way to make your days more pleasant and attract more opportunities. This can be as simple as deciding not to engage in gossip with a snarky colleague. One young investment banker I worked with told me that his goal was to be a positive presence among his peers, who worked very long hours. "I want to be the person no one minds working late with," he said. Can't you picture that kind of person? I know that's someone I'd want to work with and promote for sure.

5. Being Passionate

In what ways can you seek out opportunities to feel excited about your work—maybe not for all of your daily tasks, but at least for some of them? Where can you find even a small aspect of an otherwise unfulfilling job that you enjoy and slowly start expanding on that interest, either as part of your existing responsibilities or on the side?

In June 2020, *Harvard Business Review* featured the story of Peter, a director of HR in a Fortune 500 company. He was a top performer, yet he kept missing out on a promotion because there was no room for him to advance at his company. Although Peter liked his job and employer, he was impatient and frustrated with his lack of career growth. He was passionate about HR issues, so he began blogging about his industry on social media in his free time. A few months later, he started a podcast with a fellow HR expert. Because he also remained a top performer in his job, Peter's boss supported his extracurricular writing and podcasting and shared his posts and episodes with senior leaders. Whether Peter ultimately received the promotion he wanted or eventually left his organization, it's clear he was able to find more career fulfillment and elevate his value as a professional by tapping into his passion.

If, unlike Peter, you see absolutely no chance of liking any aspect of your current job, maybe you can find passion outside of your role and within your organization as a whole, such as volunteering to plan the annual Take Our Daughters and Sons to Work Day, or leading the Asian American employee resource group, or spearheading a weekly compost collection, or playing on the company softball team. Sometimes finding an outlet for your creativity and interests can cast a more positive light on the rest of your activities and interactions at work.

6. Using Strong Body Language

It might seem counterintuitive, but as professional interactions became virtual in the wake of the pandemic, body language actually increased in importance. Because our attention was so focused on squares of people's faces in video calls, every gesture, slouch, head nod, and eye movement became more pronounced. Some people found this to be stressful; others embraced it as an opportunity.

As suggested for job interviews, the best way to discover if your body language needs improvement is to videotape yourself during a call or presentation. How is your posture? Do you gesture too broadly or appear awkwardly statue-like? Do you rub your head when you're nervous? Do you command your space with confidence? All of this matters in person and virtually; your body language is a reflection of you and your professional reputation, and it's something you can practice and improve on, little by little, every day in any job or organization.

7. Being Coachable

How do you handle feedback when you receive it? From whom could you request more input to help you improve and grow? Remember that not all coaching has to come from your direct boss. I'm a big fan of "micro-mentoring," which refers to short pieces of advice that can come from any source. No matter your professional function or level, you can receive training, education, and guidance from a variety of places within your current organization—through formal training classes, informational resources, observing or assisting more experienced colleagues, and more.

But the magic is not in *receiving* advice or constructive feedback; the magic happens when you actually *implement* the coaching you receive. If you don't feel that you get enough coaching, then start to ask people for more (remember recalculator rule #5: ask for help). One trick is to ask for "advice" rather than "feedback." That's because "feedback" can sound like a lot of work—it implies filling out a lengthy 360-degree evaluation or giving someone an annual review. The word "advice" has a lighter touch.

If you can't think of anything specific to seek coaching for, then just get into the habit of asking people for their personal best practices on anything that's relevant to your job—technology hacks, Excel shortcuts, time-management techniques, negotiation tricks, and so on. Keep a file or notebook of what you learn. You'll benefit from the knowledge itself and from the act of inviting colleagues to show off their expertise.

To make sure that people know you're coachable—which is as important early in your career, when you're just starting out, as it is later in your career, when you don't want to be perceived as inflexible—you have to report back to them about the change you've implemented thanks to their advice. Laura Vanderkam, the time-management expert, calls this "thanking people twice," and it's a great strategy to build a positive reputation. Thank advice givers at the moment they offer you a suggestion, and then thank them again when you have taken the action they recommended. People like to know that their guidance is appreciated, and this strategy will build your reputation as a grateful, coachable, and growth-minded colleague.

8. Doing a Little Extra

"It's never crowded along the extra mile," said motivational author and speaker Wayne Dyer. One of the strategies that helped me to stand out and get promoted in my first job at WorkingWoman.com was a "little extra" habit that led to a big break. At the end of every day, I would stop by my boss's office to ask her if she needed anything before I left. Most of the time she would say no and wish me a good night. But on one occasion when I did my nightly check-in, she told me she had committed to a meeting that evening with a prominent women's organization and was too tired to attend. Would I go in her place? I said yes immediately, she called to let them know, and I enjoyed a golden opportunity to step into a more prominent role and make important connections. That was the first of many higher-level meetings she invited me to attend, which helped to elevate my profile inside Working-Woman.com and more broadly in the industry.

Elizabeth Richards, who recalculated from the field of special libraries to an operations role at a small software company, credits her success with always going the extra mile. She points out that this is especially important if you work in a smaller organization where "it's always all hands on deck for whatever needs to be done." Elizabeth advises her fellow recalculators not to limit themselves to a narrow role. "If a problem shows up, find a way to fix it or find the right person to fix it. Don't gather around looking at a problem as if it were a dead skunk that no one wants to touch. If you limit your role by thinking, 'That's not my job,' then you are also limiting how people see you. If you properly dispose of the dead skunk, you build a reputation as a problem solver."

9. Being Prepared

When I was a student, I usually did my homework. On the occasions I didn't complete the assigned reading for a given class, I hated the resulting feeling in the pit of my stomach. I was scared to be called on by the teacher and I missed out on much of the information being taught because I was busy trying to catch up. A lack of preparation in the workplace is just as problematic. If you haven't prepared for a meeting or researched a potential client before a sales pitch, you'll never be able to perform at your best.

Sometimes the smartest, most senior people are the least prepared, because they feel they can coast on their experience. One of my favorite stories of the value of preparation came from Susan, a female executive who served on a Fortune 500 company board of directors at a very young age in the 1980s. She used preparation to differentiate herself. Being several decades younger than the other board members, not to mention the only female, she felt intimidated before attending her first board meeting. She read through every document twice, made notes, and annotated her files in preparation. During the meeting she remained quiet, until a question arose that no one seemed to know the answer to.

Looking at the notes in her thick packet of materials, Susan quietly spoke up and pointed to a page. "The answer is right here in the packet," she said.

The other board members stared at her, then burst out laughing.

"None of us read the packet that carefully anymore!" one said.

Susan instantly gained credibility, thanks to her thorough preparation.

Now, I have to be honest that I find this story a little unnerving (shouldn't all corporate board members be reading the whole packet?), but it's a great lesson for recalculators: preparation can make up for a lack of experience or tenure and help you stand out in any role.

10. Having a Strong Work Ethic

Work ethic is another powerful differentiator. It is cited over and over by people who have succeeded in their careers despite growing up in a low-income community, or being the first in their family to go to college, or facing racial discrimination, or encountering any other hardship or systemic bias. While there is no excuse for bias or discrimination, remember recalculator rule #3 again: control what you can. Your work ethic is entirely in your control. If you have time limitations or physical limitations, work as hard as you can within those boundaries. Do not give up on any opportunity until you know you've put in your very best effort.

Exercise: Beef Up the Basics

To accelerate your success in any job situation at any stage, try addressing each of the ten areas above. They don't require any additional education, investment, or connections; all they require is your commitment.

1. Being on Time

What is an upcoming event for which you want to be on time or early? What action will you take to ensure you do so?

_____

_____

_____

2. Making an Effort

Identify a professional situation where you can put in more of an effort. What will you do differently?

_____

_____

_____

3. Being High Energy

At what time of day do you tend to have the highest level of energy? What is one task or activity you can move to that time of day to be happier or more productive at work?

_____

_____

_____

4. Having a Positive Attitude

How can you take a few new positive actions? For example, you can try hanging a mirror in front of your computer to make sure you smile when you answer the phone or log on to a video call. Or commit to stopping or reducing behavior that projects a negative attitude, such as sending critical texts about a colleague.

_____

_____

_____

5. Being Passionate

What is an aspect of your career, industry, or personal interests that you feel passionate about? What is one way you can add more of that activity or interest to your work, even in a small way?

_____

_____

_____

6. Using Strong Body Language

Take a video of yourself giving a thirty-second presentation. What do you notice about your posture, facial expressions, gestures, voice, or eye movements? What can you improve to be a better presenter and make a better impression on people you interact with?

_____

_____

_____

7. Being Coachable

Take a moment to ask three professional contacts for some actionable advice on any aspect of your job that you'd like to improve. It can be something small like how to respond to a particular client's email or something major like what actions you can take to land a promotion. What is the best advice you received? Write it down, then add the action(s) to your calendar or to-do list, and thank the advice givers twice.

_____

_____

_____

8. Doing a Little Extra

What is one way you can go above and beyond in your job? Can you ask colleagues if they need help on a big project? Comment on the LinkedIn post of an intern you've mentored? Proofread your big presentation one more time?

_____

_____

_____

9. Being Prepared

What are three changes you anticipate seeing in your organization or industry within the next twelve months? What can you do now to be more prepared for them (example: read an article, sign up for a webinar, subscribe to an industry newsletter)?

_____

_____

_____

10. Having a Strong Work Ethic

While not everything is in your control, your work ethic is. Is there an aspect of your job where you know you could be trying harder? Is there more you could do to succeed, but you just haven't made the effort? Write down one action you will take this week to work a little harder than before.

_____

_____

_____

Every Job Can Build Your Network and Enhance Your People Prowess

If I have one major regret in my career, it's that I didn't spend more time working for other people before I launched out on my own. I had to learn a lot of things through trial and error, things an employer could have trained me on or things I could have learned by observing and interacting with a wider variety of people—and their personalities and styles—within a professional setting. When I interview people who have worked for large organizations and received a lot of training and apprenticeship, they've spoken about learning how to act from people they admired *and* about learning how not to act from people they didn't like working with. Both are incredibly important.

Here are some relationship skills you can develop from any job or organization:

Manage Up: Learn How to Work for Different Bosses

Yes, it is your boss's job to manage you as an employee, but it is equally your job to learn how to work for, and adapt to, each person you report to. It's called "managing up," and it's an important strategy to get what you want from your career. My advice is to treat each boss you encounter as if that person is your client—supporting, respecting, helping, and aligning yourself to this person's style and metrics of success.

Here are some suggestions for how to manage your manager:

Truly understand your boss's job and priorities.

Do you know exactly what your boss does every day? What meetings do they regularly participate in? What metrics are they judged on? Whom do they report to and what are those relationships like? What pressures are they under? What are their career goals? While you can't always know the answers to all of these questions, you can learn as much as you can by getting to know your boss and using your powers of observation. The more you can align with your manager's upward trajectory, the more you can positively influence your own career path.

Know what, when, where, and how your manager likes to communicate.

Harvard Business School professor Michael Watkins, author of *The First 90 Days*, calls this having the "style conversation," and it's a game changer for working more effectively with any boss, or anyone else for that matter. Remember that the way you like to communicate may not be the same way your boss likes to receive information. To be most effective at managing up, you should be able to answer the following questions:

- Does your boss's communication style lean more toward the informal or the formal?
- About how often is it acceptable to "pop in" to their office or, if working virtually, message or call them?
- In general, does your boss prefer to receive long emails covering a lot of topics or shorter individual emails for individual topics?
- If you're running late or taking a sick day, how would your boss like to be alerted (phone, email, text)?
- Does your boss prefer to be kept in the loop on everything

you're working on (e.g., with daily or weekly update emails) or are they more hands off?

- Does your boss have any pet peeves that you should avoid doing, such as overusing the word "like" or sending emails with the message entirely in the subject line?

You can find out the above information through trial and error, consulting with colleagues, or asking your boss directly. Even if you've been reporting to someone for years, simple communication tweaks can make a big difference to your daily interactions and overall relationship.

Bring solutions, not problems.

You are going to make mistakes, so decide in advance how you are going to handle them. My best advice when facing any challenge large or small is to always bring your manager ideas, solutions, and research instead of questions, problems, and complaints. The goal is to help make your boss's job easier by reporting the facts, giving your suggestions, and asking your supervisor to decide on a course of action. In any situation, ask yourself, "What can I do here to make my manager's job easier?" The result is that your boss will see you as a problem solver, a doer, and a person they're glad to have on the team and, ideally, as a person they'd be happy to recommend to other bosses in the future.

Figure Out How to Work for Difficult People

If all of the above sounds lovely but impossible because you are currently in a role where you struggle to get along with your boss, I have a message for you: Congratulations! Experiencing difficult people, especially early in your career, is genuinely a gift. (I know from experience!) The sooner you learn how to deal with jerks, the better. This is because it is absolutely impossible to make it through an entire career without encountering any problematic people. It's a tremendous asset to your career to be known as a person who can handle a wide variety of personalities.

Bad bosses can take a variety of forms, from micromanagers, to workaholics, to criticizers, to the overly friendly, prying, personal types. (Of course, remember that you should never have to endure a truly toxic relationship, such as a boss who is verbally abusive, discriminatory, or inappropriate.)

No relationship will be as impactful on your success in a particular job as the one with your direct manager, so you'll have to figure out some coping strategies for any or all of the specific problematic types—except abusive ones, from whom you should flee. The good news is the more you build this muscle, the more confidence you'll have in handling troublesome bosses in the future. My favorite resources for specific tips are Robert I. Sutton's books *The No Asshole Rule* and *The Asshole Survival Guide*. And here are some encouraging words from recalculators who learned some good lessons from some bad managers:

> I once had a very negative boss. As a woman of color and a subordinate, and also a single mom with two children afraid of being let go, I was afraid to challenge her. I just continued to produce strong results and went out of my way to kill my

boss with kindness, while collaborating with other departments, her peers, and employees above her level. After a while, the bad boss stopped bothering me because I made her look good. I know now when dealing with a "bad" boss to maintain your strong performance, exhibit the behavior and professionalism that you expect, and align yourself with cheerleaders. I built a support system and network that was bigger than that boss.

–*Rhea Faniel, senior associate director, diversity recruitment and employer relations, City College of New York*

From my worst bosses, I've learned the value of autonomy and respect. I know how I feel when a boss refuses to let you grow or experiment with new ways of working and thus, I try never to do that to those who chose to work with me on my teams.

—*Jennifer Owens, former editorial director,* Working Mother *magazine; currently founder and president of Jennwork, a content agency*

I once had a supervisor who was not present, lacked all detail orientation, and failed to recognize his employees. As a supervisor and leader, I've learned from this experience and try to provide my team with supervision, feedback, direction, and encouragement.

—*Jason Eckert, executive director of career services, University of Dayton*

Keep Building Your Contacts

Every job offers an opportunity to build your professional connections and expand your network. Even if you change industries in the future, you never know how your current contacts might be connected to people in other fields. I've often found that the happier people are in an organization, particularly a large one, the less likely they are to network because they feel secure. Given the realities of the twenty-first-century economy, I don't think it's ever safe to believe that one's network is somehow "complete."

Many years ago I met a financial services professional and asked if her firm ever brought in speakers like me to conduct workshops on LinkedIn or professional networking. "We don't need that," she told me. "We have strong internal networking and development programs so people don't need to network outside our firm." That firm was Lehman Brothers, which went bankrupt overnight at the beginning of the 2008 global financial crisis, leaving many professionals without outside connections when they needed them most. You *always* need a network.

Here are some ways to build more contacts while working in almost any job or industry:

- Sign up for any formal mentoring or networking programs offered by your organization, such as mentor matching or speed-networking events.
- Join a committee or attend an online or in-person event hosted by an employee resource group (ERG, also known as an affinity group), which are internal, employee-led networks that exist to bring people with similar identities

or interests together, often with the goal of increasing diversity, equity, and inclusion. Depending on your organization, you might find groups for Black employees, Asian American employees, Latinx employees, LGBTQ+ employees, veterans, people with different abilities, women, multigenerational employees, working parents, and more. People who identify as allies or supporters are often encouraged to join as well.

- Volunteer for a charitable activity hosted by your employer. Volunteering is one of the easiest and most effective ways to meet a variety of colleagues across job functions who share similar interests to yours.

- Participate in your company's LinkedIn group, perhaps promoting posts from internal recruiters to your network and sharing announcements about the organization. Connect with coworkers and comment on their posts to demonstrate support.

- When you are invited to attend any social gathering, make it a habit to default to yes. This is a great way to expand beyond your current group of work friends and colleagues, and it's my rule of thumb when I'm trying to increase my business or raise my profile. You might join the welcome video call for interns who are significantly junior to you—they might be on your team someday (or might hire you—junior people move up fast!). Attend the retirement party of a colleague you've never worked with—you never know whom you might strike up a conversation with. Log on—camera on—for the online presentation of a colleague in India, even if you might never meet them—you might gain an insight that could help spark an idea for your team. If you want to meet new people and achieve new things, you have to place yourself in new situations.

If you're struggling to find opportunities to network in your organization or industry, maybe it's time to create your own group. At Penn State University's Lehigh Valley campus, a group of administrative assistants wanted to network and develop in their careers but couldn't find a group to join. They decided to create their own network, called Administrative SuperPowers, through which they brought in professional development speakers and created workshops for members to demonstrate their own expertise to the broader campus community. For example, when the campus switched to Microsoft 365, the Administrative SuperPowers hosted an event where they offered tips and tricks on the new software. One member of the group impressed higher-ups so much that she was promoted to a more senior role on campus.

Every Job Can Improve Your Self-Awareness, Personal Growth, and Work/Life Integration

Another way to turn any job into a great job is to use it as an opportunity for self-reflection and personal growth. What can you discover about yourself through your current role? What opportunities can you find to improve your skills of self-advocacy? Can you learn to manage your time or your stress better in this position?

One of the biggest and most positive changes to work that I've observed over the past twenty years is the increased attention to work/life integration issues among professionals of all ages and personal situations. That focus only intensified during the pandemic, when so many of us had no choice but to live, work, and manage our personal relationships all under one roof.

A new job or a recommitment to an existing job offers the perfect opportunity to find the work/life mix that works for you. Here are some tweaks to consider:

Block Your Time.

Whether it's working remotely (either part-time or permanently) or taking small breaks throughout the day, increased flexibility is often a determining factor in better work/life integration. One technique that became more widespread as many people reevaluated their schedules during the pandemic was "windowed work" or "time blocking," which is the ability to break up your day into distinct segments of work and personal time. For example, you might switch between two-hour work blocks and personal blocks throughout the day, adding up to a full eight hours of work, but not straight through from 9:00 A.M. to 5:00 P.M.

In a 2020 survey from Robert Half, 78 percent of respondents with kids said the practice made them more productive, as did 66 percent of those without kids. Take the chance as a recalculator to set up work habits that work best for you.

Create a Schedule That Adapts to Your Peak Times.

Being an early bird or a night owl is a thing . . . and it's related to your circadian rhythms. Some people are most productive at 7:00 A.M., while others fire up at 7:00 P.M. While you can't necessarily make everyone on your team work around your schedule, it's worth noting your preferences and talking to your manager about how you can adapt to your own power hours, especially if you are part of a remote or hybrid team with people in multiple time zones. One smart practice is to clarify time-related deadlines, such as "end of day" for deliverables. Is that your boss's end of day, or would it be fine if you sent it off at your end of day—even if that's 11:00 P.M.— just as long as they have it by first thing the next morning?

Just Say "No" (Politely).

Ultimately, much of your ability to improve your work/life integration comes down to setting boundaries and avoiding the dreaded "scope creep," that phenomenon when one task seems to morph into five. If you're a recalculator, you might feel the need to work extra hard to prove yourself, but please don't let this lead to burnout. A new manager or job presents the optimal time to start fresh by asserting what you need to be your most productive, happy self at work—and often that means removing something from your plate before it ever gets comfortable there.

Every Job Can Lead to the Next One

Remember back in the Introduction when I talked about the fact that recalculation is a skill you'll be tapping into for the rest of your career? This is certainly true in situations that are not ideal, but it also applies to professional situations in which you are quite happy. Given all of the unprecedented disruption to our world in recent years, I don't think it's ever safe to rest on your laurels. HR expert and career consultant Tashana (TD) Sims-Campbell wisely says, "You should never get too complacent, and it's easy to do, especially when you love what you do and where you work. We should always prepare ourselves for the next level, whatever that looks like and wherever that is."

Here are some specific strategies to consider:

Bake Your PIE

If you've worked for a large organization in the past three decades, you may have come across the "pie chart of success in high performing organizations" from Harvey J. Coleman's book, *Empowering Yourself: The Organizational Game Revealed*. It is an actual pie chart divided into three sections: "**P**erformance" represents about 10 percent of the chart, "**I**mage" represents about 30 percent, and "**E**xposure" represents about 60 percent.

The message of the chart is not that performance doesn't matter; of course, performance in one's job is critical. The point is that for many professional jobs today, doing your actual work is the baseline for success—in business, it's table stakes, or a minimum entry requirement. Throughout this chapter I've worked from the assumption that you are doing the actual work of your current job well. You are getting things done and meeting expectations. But if you work in an organization where most people do their work, what matters more to your success, then, is the combination of your image and exposure. Yes, we're talking about personal brand and your career story again. They are just as important when you are happily employed as when you are job hunting.

The question to ask yourself is this: Do other people *know* about you and your good work? One of the biggest mistakes people make is to assume that someone will notice their great performance and all they need to do is keep their head down and get things done. Nope! You have to make sure your colleagues and higher-ups know the value you bring. This is even more vital when you are working remotely or on a virtual team or in a company or industry that is under pressure from market forces. No one else is going to promote your excellence—that responsibility is in your hands.

Here are some suggestions for increasing your exposure and bettering your image in any job, to help you get tapped for advancement opportunities:

- When you have a win at work—such as finishing a big project, signing a new client, finding a way to save money for your employer, nailing a big presentation, or receiving a compliment from a customer—don't be shy about sharing the news with people senior to you. This is particularly important if you work remotely. Keep a running list of your positive accomplishments, which you can refer to during midyear and/or annual reviews, and quantify when possible. Better yet, it is totally appropriate to send a brief email update reporting your positive results immediately after a win. You can also forward positive messages from clients or colleagues to your manager. Give credit to colleagues if the win was a team effort, but don't be too humble about your role in any success. (P.S. Add the biggest achievements to your résumé and LinkedIn profile. Don't wait until you're job hunting again to try to remember what you've achieved.) Again, bear in mind that no one else is keeping track of your biggest accomplishments; it's your job. If you don't toot your own horn, no one else will.

- Make a point to contribute on all phone and video calls and at in-person meetings. If this makes you nervous, plan in advance by jotting down a few notes. If you wait to be "called on" to speak, you might be waiting a long time.

- Volunteer for new projects and cross-functional opportunities where you can work with teams that you don't usually interact with or participate in committees that will offer you increased exposure. Some larger organizations have "opportunity marketplaces" where such projects are posted. These additional responsibilities can require more work, so be selective about which ones will have the most impact. One of the advantages of a more virtual work environment is that it's easier than ever to collaborate with colleagues in different locations and on different teams.

Consider Internal Mobility

If you like your current employer but don't love your current role, one of the paths to consider is an internal move. When you've increased your image and exposure enough, you might even be offered these opportunities. Internal mobility can be tricky to navigate—you don't want to offend your current team or step on any toes with the team you might want to move into—but it can be a positive and satisfying recalculation. At many large global corporations with multiple divisions, an internal move can even feel like transitioning to an entirely new organization.

To explore such a move, Dan Black of EY suggests reaching out to your human resources department and, first and foremost, making it completely clear that you are not abandoning your current role. He suggests a conversation in which you say something like, "I am committed to the organization and committed to excelling in my current role. But I am interested in doing some exploration in X area. Are there any opportunities where I can get some exposure to that part of the organization, perhaps by doing a shadow day to observe someone in that role or by pursuing some additional training on my own time?"

Dan emphasizes that you'll likely have to be flexible when it comes to your timeline for potentially making an internal move. "The company's timeline is likely not the same as the timeline you have in your own head," he points out. But with some patience and hard work, internal moves can be rewarding for both you and your company, which doesn't want to lose great talent.

Always Leave with Grace

Whenever you do leave a team or organization, be sure to do so in a positive way, without burning bridges. The best course of action is to deliver the news of your departure by focusing on your enthusiasm about what you are heading toward rather than any complaints you have about the place you are leaving. Remember that life and careers are long and you never know who you might meet again on the road ahead.

As you can see, there are many ways to grow, reflect, pivot, advance, and yes, recalculate, through any job at any time. Zach Mercurio, the Colorado State University professor, shared a lovely parable with me that I hope you'll find as inspiring on this topic as I did:

There's a legend in the stonemasonry community about a man who happened upon three stonemasons tediously chipping away at chunks of granite. The first stonemason was unhappy, looking at his watch and scowling. The man asked the mason what he was doing. The first mason said, "What does it look like? Just hammering this stupid rock so I can get home at 5:00 P.M. and get paid."

The second mason was more into his work, hammering more carefully. When the man asked this mason what he was doing, he said, "I'm molding this rock so it can be used to finish that wall over there. It's not too bad, but I can't wait until it's done."

The third mason was hammering and chipping diligently and methodically, frequently stopping and running his hand over the rock and stepping back to admire his work. The first time the man asked him what he was doing, he didn't hear him because he was so engrossed in the task.

Finally, this third mason said, looking skyward, "I'm building a cathedral!"

7

Move On and Move Up

Meet Real-Life Recalculators Who Have Put the Lessons of This Book into Action

> Every success story is a tale of constant
> adaptation, revision, and change.
> —RICHARD BRANSON

Early in my career, I attended a conference where top business leaders shared the wisdom they most wished they had known when they first started out. I'll never forget the words of one executive, who said, "Looking back on my career, I wish I had told my earlier self not to be so afraid."

She went on to share that her biggest regrets were always about actions she *didn't* take, jobs she *didn't* apply for, and ideas she *didn't* pursue, rather than any negative outcomes from risks she actually took. That advice has stayed with me to this day, and I frequently put it into practice when I'm hesitating to try something new.

Asking people, "What do you wish you had known when

you were in my situation?" remains one of my favorite ways to learn from people who've "been there, done that" before I have. Sometimes one comment, one story, one detail can make an enormous impact.

That's why my parting advice to you is this:

Recalculate forever.

If there is one overarching lesson that a global pandemic has taught all of us, it's that you never, ever know what lies ahead.

Even when you're feeling settled in your career, remain open to opportunities. Even when you are at the height of your field, keep learning. Even though you know a ton of people, continue expanding your network. Even when you're busy, try something new. My husband, Evan, says this is the best advice he was ever given by a mentor: *Always take the meeting. Always.* You just never know.

In this final chapter, you'll meet a variety of successful recalculators. Their stories confirm the truest of clichés: that the only constant in life is change. Industries rise and fall, economies boom and bust, health declines and improves, relationships begin and end, and whenever you think you've "arrived" at a particular life or career destination, another starts coming into view. The choice is yours whether to resist this unpredictability or to embrace it.

The Mindset Recalculators

While life can change on a dime at any moment, the one thing you can always control is your attitude. While it is healthy to feel disappointment, sadness, anger, or any other emotion when circumstances are tough, you can heal, recover, pivot, and find opportunity.

The good news is that there is a lot of precedent for finding opportunity in times of crisis and change. Did you know that General Motors, IBM, Disney, Microsoft, Electronic Arts, and Trader Joe's were all founded during economic recessions? It was worth experiencing a bad economy if the result was Everything But the Bagel Sesame Seasoning and *The Little Mermaid*, no?

Here are other real-life examples of recalculators who adjusted their mindset to find career fulfillment in times of change:

There Are No Wrong Choices.

If you say no to opportunities because they aren't listed in your five-year plan, you might just be saying no to your destiny. Conversely, if you can't recognize when something isn't working, you're doing yourself an equal injustice. . . . I see doors and windows open daily. I ignore some, I walk through others. There is no wrong way to go. There's just different. Success is making things happen that you never thought you could. It's also being able to realize when something you thought you wanted isn't something you want after all. We get one life, but many chances. It's up to us to take them.

—*Aliza Licht, New York, NY*

Attitude Is Everything.

[After graduating college into the global financial crisis in 2008], I really just enjoyed whatever situation I was in at the moment. If you had told me two years before I graduated that I'd be working at a grocery store and loving it, I would have thought you were ridiculous. But I did love it. I enjoyed every moment of it. And I took the opportunity to learn as much as I could while I was there.

—*Eliza Parsons, Seattle, WA*

You Might Surprise Yourself.

Do not think too far in advance. Do not be set in stone with the idea of what you think your life should look like. During my two years in college so far, every time I think I know where I'll end up in the next year, it turns out to be 100 percent the opposite. I remember thinking, "I will never be an executive," and now I'm in a business fraternity!

—*Hannah White, Columbia, SC*

The Pathfinding Recalculators

Sometimes you'll land in a spot that just feels wrong. Maybe you're disillusioned with your current job or industry, or you've relocated to a city you don't like, or you've started a business and realize you don't enjoy self-employment. It happens. The key to moving on is to know that every perceived "mistake" or dissatisfaction is an important learning experience to push you toward a better path. There are truly no mistakes as long as you learn from them. Here are some tales from recalculators whose careers have been anything but a straight line:

You're Allowed to Change Your Mind.

I was feeling burnt out from many years of toxic corporate life, and became interested in pursuing a life of learning through a master's degree in library and information science from one of the best universities in that space. I made my case to the dean and was accepted into the program. Even though I gained a number of search skills that I still use, I concluded that the cost/benefit ratio, long-term, was not in favor of this new path. I did, however, use those same search skills to position myself for higher-level jobs, and succeeded in landing one the following year.

—*Andy O'Hearn, Bridgewater, NJ*

It's Never Too Late to Find Something New.

After nineteen-plus years as a professional at a well-known telecommunications company, I was "downsized," i.e., laid off. At that point in my career and then a mother of two, I never thought about leaving; I was a lifer. I had benefits and a good salary. It did not matter that I had been getting bored and not feeling challenged and that one of my peers was a constant source of drama for a relatively strong team. While it took more than two years of consulting, working grant-funded and temporary jobs, and a stint as a sales associate at a department store, I landed what I did not know at the time would be my dream job that I still love almost sixteen years later.

—*Rhea Faniel, New York*

You Can Trust Your Instincts.

After I graduated from college, I decided to prioritize a personal relationship over my professional career. My now-husband entered the Army after we both graduated from college, so I traveled with him. For the next five years, I questioned myself. I was happy, and I was pursuing my master's degree online, but I couldn't help but question whether I had made the "right" choice. At that time, I considered this to be a detour in my plans to become a "successful professional." I often reflect on this choice and how much it impacted my development of having trust in my decision making. I'm proud to have pursued a detour rather than the safe path. It reminds me of the Robert Frost quote, "I took the [road] less traveled by, and that has made all the difference."

—*Cheryl Rotyliano, Ithaca, NY*

Your Passion Might Not Be What You Thought It Was.

During my undergraduate career, I started an internship for a media production company. After completing the internship I was offered a full-time role during my junior year in college. I thought this meant all of my dreams were coming true. However, after working there a couple of months full-time and crying daily because of the stress, I realized that career path wasn't for me and had to quit. This then led me to thinking about other options and going to graduate school. During my first semester of graduate school, I found the world of career services in higher education helping other students find their career paths. I now know my passion because I had that detour.

—*Matthew French, Charlotte, NC*

Failures Can Point You in a Better Direction.

I made the last-minute decision to attend graduate school for an MBA so I could get experience in sports media relations through an assistantship that would help me to become a sportswriter. After a failed accounting class, I switched from an MBA to a master's in counseling. When it came time to choose a practicum, I found one at a student-athlete academic services office at a school two and a half hours away. Somewhere during that practicum, I found my new career path. I realized that what I enjoyed most about sports writing was the story behind the athlete. That was twelve years ago and I am still working in the student-athlete services profession today.

—*Lynaye Stone, Williamsburg, VA*

The Career Story Recalculators

What story are you telling yourself about your career? Your professional reputation begins with what you believe to be true about yourself. Make sure you are crafting your narrative and personal brand with yourself cast as the protagonist! Here are examples of people who've done just that:

Your Career Story Can Include Both Making Money *and* Having the Lifestyle You Want.

When my husband accepted a position out of state, I knew I would be leaving my full-time job. With childcare being so expensive in our new city, we decided I would stay home and try to start a résumé-writing business, Get Your Best Résumé, to earn some income as well. This decision was a detour that could not have gone better. My little side hustle became a full-time job where I began making the same level of income I did when I was working full-time. I also had the flexibility of managing my own schedule.

—*Sara Hutchison, Nashville, TN*

Even in Times of Uncertainty and Transition, You Can Enhance Your Career Story.

I came home May 23, 2020, right after my virtual college graduation from UC Berkeley. My first round of medical school applications were due May 28. I grinded to get my primary applications done, then figured I would take a break at home with my parents in Los Angeles until my secondaries were due in July. My full-time job offer had been postponed for six months, so I thought I would hang out until the job started or just wait for med school. That was, until I had an influential phone call with my friend's dad, who is a doctor. He said, "You're missing the opportunity of a lifetime! Your medical school interviews are going to ask what you did during COVID-19 and you can't tell them you did nothing." I knew he was right, but I was afraid to take an in-person job and possibly pose a health risk to my parents. He recommended that I look for a virtual contact tracer position, which they were hiring in droves. That idea completely shifted my mindset and compelled me to find something meaningful to do with my time. I applied for contact tracer positions and ultimately got a related job for a public health research group funded by the CDC working to help track the epidemic.

—*Olivia Goodman, Berkeley, CA*

The Networking Recalculators

You can't achieve anything in life entirely by yourself. This means that the people you surround yourself with are the co-stars of your career journey. As Rob Lowe (or, as I will forever think of him, Sam Seaborn on *The West Wing*) wrote in his memoir *Love Life*, "I think it was Alfred Hitchcock who said that 90 percent of successful movie making is in the casting. The same is true in life. Who you are exposed to, who you choose to surround yourself with, is a unique variable in all of our experiences, and it is hugely important in making us who we are." If you want to pivot your career path, keep the supporters, move on from the non-supporters, and build relationships with new people in the realm you want to join.

Never Let Go of Your Network.

When my mom got diagnosed with cancer, I was thirty and finally was in a job that I really liked. I made the decision to quit that job to be with her and take care of her. Although my career was starting to come together, I knew nothing was as important as being with her during the last months of her life. I was so uncertain about the future and wondered if I would ever get back on course. After she passed away, I transitioned into the phase of life where I had children and didn't work outside of the home. I was always a little anxious about being out of the workforce for so long. During that time, which lasted approximately six years, I made sure that I kept my credentials current, stayed connected with colleagues, actively participated in professional organizations, and engaged in relevant volunteer work. Because of my strong professional relationships, I was able to land a great job when the time was right again. Now, many years later, I just got divorced and I'm glad I am comfortable with change! It's so important to always have options, even if you never need to use them.

—*Jessica Scott, Dallas, TX*

One Conversation Can Change Your Entire Trajectory.

The economic downturn in 1983 in Houston led to the loss of my job as operations manager for a new luxury hotel property. The entire industry was depressed and in a major regional recession. I had just relocated to Houston from Dallas, where I had left behind a relationship that mattered to me. A computer systems vendor who provided products I had previously purchased for multiple hotels told me, "If you ever wanted to, you could sell these systems and I'm sure you'd do well. And you could do this while located in any major U.S. city." That comment inspired me to move back to Dallas and embark on what turned out to be a successful twenty-five-year career in sales—and the start of my current, thirty-six years and counting, wonderful marriage. Eventually, the combination of my experience in technology, sales, and hospitality led to my final position—my dream job—working for Hilton.

—*Ed Roach, Dallas, TX*

The Job Search Recalculators

Job searching can be a stressful experience, but it can also be exhilarating. Every job description you encounter represents a door to an entirely new path. Don't unnecessarily limit yourself by only applying to jobs with a certain title, a certain type of organization, or a certain industry. Even small adjustments to your job search criteria can make a big difference in the opportunities you open up for yourself. Here are some stories to inspire you to think more expansively:

Don't Be Afraid to Move Laterally.

There was a moment in my career where I wanted to jump from trade magazines to consumer, but that move came with a huge stigma that resulted in me losing jobs to less-experienced candidates, more than once. I'd never experienced this before, so it was a huge shock. I kept pushing, and when an opportunity arose to lead a group of craft magazines, I moved laterally for the first time in my career, and that made all the difference in creating a new path for me. Since then, I've seen the value of following opportunity rather than the expected path.

—*Jennifer Owens, Brooklyn, NY*

It's Okay to Start with a Low-Paying Job to Get Your Foot in the Door.

Don't be afraid to take a low-paying job if it has the potential for growth. Because when the economy comes back, they'll hire more people, and then you're in the driver's seat there because you're the most experienced person. In my department, most of the people who are managers started before the recession, and . . . once we got to 2010, 2011, we were hiring a ton of people, and so the people on were on before the recession, if they were good enough, then they became managers. So as a younger person, if you are able to just get your foot in the door, when the economy comes back, there will be a lot of people banging on the door, but you have a leg up because you have the experience.

—*Tony Trepanier, Seattle, WA*

The On-the-Job Recalculators

The Russian actor and director Konstantin Stanislavski famously said, "There are no small parts, only small actors." I feel the same way about jobs: there are no inherently "bad" jobs. You can turn absolutely any job into a valuable stop on your journey. Maybe because you learn a new skill, or meet a key person, or even just discover that you never ever want to do that job again. Here is some inspiration:

Every Job Will Teach You Something.

When I first graduated from college with a journalism degree, I did a lot of freelance writing work. After a few years I started to feel disillusioned with the lifestyle and applied like crazy trying to find something more consistent. A friend was working at a market research consulting firm and offered me a job in the data department, so I took it. I fell into a slump working there for *much* longer than I should have—four and a half years! Although I enjoyed collaborating with my coworkers, it wasn't what I wanted to do. I kept thinking, "This job is so pointless, what am I really getting from this?" But when I finally landed a job in my dream industry of book publishing, I would find myself catching little details I wouldn't have noticed before. I've always been detail oriented, but my data job brought that skill to a completely new level. I also became much more proficient with workplace productivity tools. You might think you are in limbo, but there is always going to be something to take away from any position.

—*Wendy Wong, New York, NY (Note: Wendy is the editor of this book!)*

Taking the "Wrong" Job Can Pay Off.

I was floundering in my career and had just returned from two and a half years living in Liberia, first as a trailing spouse and then working at the U.S. embassy. For about nine months, I made every effort to find a position in my preferred field and completed several short-term contracts. Eventually, I was offered a managerial position back in Liberia in a field outside my area of interest. Despite my gut telling me this was not the job for me, I took it. The next nine months were the hardest of my life on many levels. At the end of that experience, though, I was offered a leadership position in my area of interest that ultimately led to better things. Although it took more than four years, I got on the right track and know *a lot* more about myself.

—*Paul Binkley, Fredericksburg, VA*

You Can Overcome Mistakes and Failures—Even Really Big Ones.

Due to a significant and regretful mistake on my part, I was given the sad choice to leave or be terminated from my position at a state university, where I had been a department chair and full, tenured professor after eighteen years. Luckily, I was able to attain a dean's position at a community college in the state, and that position was changed during the year to a vice president position. All I ever wanted to be was a professor; I've been in higher ed for thirty-five years and have a passion for teaching. But being more or less thrust into administration, I've learned an entirely different set of skills that I didn't know I had. Though I didn't "recalculate" my career by choice, I've had the opportunities to recover and grow and succeed.

   —*Dr. James Yates, El Dorado, AR*

Sometimes the Best Recalculation Involves Staying Put.

There was a job I wanted, early in my career. I could taste it. It was with a prestigious employer, and I was the front-runner. I was wined and dined. My references were checked. Word got out, and my current staff and colleagues were planning for my departure. It was mine! Or so I thought. And . . . the offer never came. Sharing this takes me back to a sinking feeling I'll never forget. But, in digging out of the disappointment, I was able to reenergize with my existing job, earn a coveted promotion, and compete successfully one year later for an opportunity that was a bigger challenge and a much better fit.

   —*Andy Ceperley, Palm Springs, CA*

When I first set out to write this book, I started to tell people about the concept and would simply ask them, "Have you ever recalculated in your career?" I was overwhelmed by the responses and truly had no idea about the number of people who had experienced recalculations, the variety of recalculations they had undergone, and the range of outcomes they achieved. The above stories are only the tip of the iceberg. Whenever you are struggling or uncertain during your recalculation journey, ask people around you about their own transitions. I guarantee you will find inspiration everywhere.

Conclusion

Everything will be okay in the end.
If it's not okay, it's not the end.
—JOHN LENNON

As I finish writing this book, I am sitting with uncertainty. What will the world be like when this book is published? What else will have changed? Will anything be back to "normal," whatever that means? I have never experienced anything like the roller coaster of 2020, and the unprecedented ride is far from over. The truth is that I've had to recalculate many times while writing this book and will continue to do so—along with all of you—in the months and years to come. The economy will continue to rise and fall, companies will open and close, technology will evolve and change, and health concerns will continue to be top of mind.

To quote my favorite tweet of 2020, "Don't know about y'all but I could really go for some precedented times."

And yet, perhaps my biggest takeaway from writing this book is that times are never really stable in the modern world. As we've explored, the pandemic merely accelerated work trends that have been building for decades. It can feel scary and challenging that the "old" rules of work no longer apply, but it's also exhilarating. Together, we have a once-in-a-lifetime opportunity to forge entirely new career paths and work schedules, create new ways of making a living, restructure work environments, take more action to make the workplace more diverse, equitable, and inclusive, and invent new ways to make work a more positive experience for more people.

I was struck by a comment my cousin Olivia made about being a member of the infamous graduating Class of 2020. "I'm envisioning job interviews for the rest of my life having an awkward pause when the interviewer notices my graduation year," she told me. I couldn't help thinking that we would all feel that way about *anything* we did in 2020 and the entire period of time dominated by the pandemic—such an unexpected, complicated, heartrending, universally shared experience. How did it shape us? What did we create from it? How did it change each of our individual paths as well as our collective future?

While these answers are still very much in progress, I want to take this final moment to wish you success, fulfillment, and good luck on your recalculation journey. Remember that we are all in this together, and you are never alone on the road.

I'll see you out there.

Lindsey

Acknowledgments

This book was born from a tiny little sentence at the end of an email about an entirely different topic. We were deep in lockdown in New York City, and the email sender was my literary agent, Michelle Wolfson:

"I think you can use this time for another book."

And here we are!

Enormous thanks to you, Michelle, for giving me a purpose during a difficult time. Thank you for your support and confidence in me over so many years. I am grateful.

Thank you to Hollis Heimbouch for embracing this book from the very beginning, and for your support over many years and books. Thank you to Wendy Wong for going above and beyond, week by week, line by line as we made this book happen. I couldn't have done it without you. Thank you to each member of the HarperCollins team for your support of me and *Recalculating*, including Brian Perrin, Laura Cole, Nicholas Davies, and Andrea Guinn. I am deeply grateful for your hard work and positivity.

For assistance with writing and editing in the midst of a pandemic, wildfires, college move-ins, and more, thank you to Cathie Ericson.

Huge gratitude to Eileen Coombes and Rebecca Slagowski for helping to keep my business running while I wrote and for your positivity and professionalism every single day. It is such a pleasure to work with you both.

Many thanks to all the subject matter experts and recalculators who made introductions and shared your insights and experiences for this book. Your stories and support are deeply appreciated.

Especially in the fraught year of 2020, I am endlessly grateful and feel extremely fortunate to have an amazing network of family, friends, mentors, clients, and colleagues who support me personally and professionally. Thank you especially to Roni Ayalla, Terri Bacow, Deb Berkman and Ben Rosenblum, Meredith Bernstein, Derek Billings, Lisa Brill, Fred Burke, Wanda Echavarria, Ilana Eck, Diana Fersko, Nicky Garcea, the Goodman/Ramsay family, the Gotlib family, Cynthia and Greg Hudson, Erica Keswin, Mignon Lawless, Jeannie Liakaris, Alex Linley, Danielle Martin, Kelly Nadel, Solana Nolfo, Andy O'Hearn, the Raho family, Amanda Schumacher, Cari Sommer, Katie Speyer, Lindsay Starr, Trudy Steinfeld, Manisha Thakor, Amy Vanderwal, Chelsea Williams, and Courtney You. Special thanks as well to everyone in the PS9 community for your above-and-beyond support this year.

Thank you to Suzi Weisberg for being a wonderful caretaker and friend—you are honorary family for sure.

Thank you to Mom, Dad, Rob, Laura, Anne, Betty, Jon, Owen, and Will for your support and love always.

And to Evan and Chloe, who cheer me on every day. I love you more than rainbow sprinkles.

Notes

Introduction

3   In October 2020: Bryan Walsh, "AI and Automation Are
     Creating a Hybrid Workforce," Axios, October 31, 2020.

3   According to the Federal Reserve: Derek Thompson, "The
     Coronavirus Is Creating a Huge, Stressful Experiment in Working
     from Home," *Atlantic*, March 13, 2020.

3   as of March 15, 2020: Megan Brenan, "U.S. Workers Discovering
     Affinity for Remote Work," Gallup, April 3, 2020.

4   After 128 months of economic expansion: Ramona
     Schindelheim, "We Have an Unemployment Crisis. Now What?"
     Jewish Journal, July 8, 2020; and Nicole Lyn Pesce, "55% of
     Businesses Closed on Yelp Have Shut Down for Good During the
     Coronavirus Pandemic," *MarketWatch*, July 22, 2020.

6   As the Future Hunters, a futurist consulting firm, affirmed:
     Future Hunters, "Future Bites: 11 Emerging and Critical Trends
     for 2020 and Beyond," 2020, https://thefuturehunters.com/our
     -work/future-bites-2020/.

9   According to research conducted by ZipRecruiter: Julia Pollak,
     "How Unemployment Benefits Affect Job Search During
     Coronavirus," ZipRecruiter blog, https://www.ziprecruiter.com
     /blog/how-unemployment-benefits-affect-job-search-during
     -coronavirus/.

Chapter 1: Adjust Your Mindset

23   Research has found that passive scrolling: David Ginsberg and
     Moira Burke, "Hard Questions: Is Spending Time on Social Media
     Bad for Us?" Facebook blog, December 15, 2017.

28   Matt Doyle, an actor in the musical: Joshua Barone, Joe
     Coscarelli, Julia Jacobs, Gia Kourlas, and Zachary Woolfe,

"New York's Arts Shutdown: The Economic Crisis in One Lost Weeekend," *New York Times,* September 23, 2020.

28   2018 Bankrate study: Amanda Dixon, "The Average Side Hustler Earns over $8K Annually," Bankrate, June 25, 2018.

34   And it seems the memo has been received: Shelley Osborne, "How to Train Workers for a Growth Mindset," *HR Dive,* July 7, 2020.

39   40 percent of Americans have been laid off or terminated: Harris Poll, "2019 CareerArc Layoff Anxiety Study," CareerArc, 2019.

39   In some instances, it can take longer to get over the grief of a layoff: Andrew J. Oswald, "Death, Happiness, and the Calculation of Compensatory Damages," University of Warwick, October 31, 2007.

44   One 2020 study found that nearly two-thirds: David Louie, "How Are Millennials, Gen-Z, Boomers Coping with COVID-19 Stress? Surprising Survey Results Here," ABC7, July 30, 2020.

Chapter 2: Forge Your New Path

52   "Begin with the end in mind": Stephen R. Covey, *The 7 Habits of Highly Effective People* (New York: Simon & Schuster; Anniversary edition, 2013).

55   Pearson found that 73 percent of people globally: "The Global Learner Survey," *Pearson,* August 2020.

55   The "career lattice" concept was conceived: Cathy Benko, Molly Anderson, and Suzanne Vickberg, "The Corporate Lattice: A Strategic Response to the Changing World of Work," *Deloitte Insights,* January 1, 2011.

56   Sheryl Sandberg, chief operating officer of Facebook: Heather McCullough and Eric Behrens, "It's a Jungle (Gym) Out There: Charting Your Career as a Lifelong Adventure," *Educause Review,* August 9, 2016.

56   Executive coach Nihar Chhaya: Nihar Chhaya, "The Upside of Career Envy," *Harvard Business Review,* June 16, 2020.

56   author and blogger Anne Bogel: Zibby Owens, host, Anne Bogel, *Don't Overthink It, Moms Don't Have Time to Read Books,* March 16, 2020.

63   Six out of ten: "7 Reasons Adult Learners Are Going Back to School," Post University blog, April 17, 2020.

63 70 percent of full-time college students: Madeline St. Amour, "Working College Students," *Inside Higher Ed*, November 18, 2019.

63 "Your children can expect to change jobs": Thomas Friedman, "After the Pandemic, a Revolution in Education and Work Awaits," *New York Times,* October 20, 2020.

64 One study in 2020: "Generation Z Considers Traditional College Path as Old School," *ECMC Group/VICE Media*, May 2020.

64 many employers in the popular tech sector: Glassdoor Team, "15 More Companies That No Longer Require a Degree—Apply Now," Glassdoor, January 10, 2020.

64 widest earnings gap between the two groups on record: Christopher S. Rugaber, "Pay Gap Between College Grads and Everyone Else at a Record," *USA Today*, January 12, 2017.

65 "Learning is the new pension": Friedman, "After the Pandemic, a Revolution in Education and Work Awaits."

65 after the first few months: Paul Fain, "Looking Beyond the College Degree," *Inside Higher Ed*, June 24, 2020.

68 Julie Lythcott-Haims, former dean of freshmen at Stanford University, who shared some unique advice: Morra Aarons-Mele, host, "How to Stop the Cycle of Overachieving," *The Anxious Achiever with Morra Aarons-Mele*, HBR Presents, June 22, 2020.

70 Even Google, which used to be famous for asking candidates of all experience levels: Adam Bryant, "In Head-Hunting, Big Data May Not Be Such a Big Deal," *New York Times*, June 19, 2013.

73 Personal finance expert Ramit Sethi reminds potential applicants: Megan Leonhardt, "Here's How to Decide If Grad School Is Worth It, According to the Author of 'I Will Teach You to Be Rich,'" CNBC, *Make It*, July 1, 2019.

Chapter 3: Your Career Story

87 she once had these wise words to say: Hilary Burns, "Morgan Stanley's Carla Harris on Why You Need to Understand the Adjectives of Success," *Bizwomen*, October 3, 2014.

88 My preferred definition of personal branding: "Definition of a Personal Brand vs. Personal Branding," PersonalBrand.com, https://personalbrand.com/definition/.

90   seminal 1997 *Fast Company* cover story: Tom Peters, "The Brand Called You," *Fast Company*, August 31, 1997.

96   We are currently in the midst of what economists have identified: Nicholas Davis, "What Is the Fourth Industrial Revolution?" World Economic Forum, January 19, 2016.

96   According to a 2018 independent task force report: Edward Alden and Laura Taylor-Kale, "The Work Ahead: Machines, Skills, and U.S. Leadership in the Twenty-First Century," Council on Foreign Relations, April 2018.

96   As Ramona Schindelheim of WorkingNation reported: Ramona Schindelheim, "We Have an Unemployment Crisis. Now What?" *Jewish Journal*, July 8, 2020.

102 A team of researchers: Michelle R. Weise, "Research: How Workers Shift from One Industry to Another," *Harvard Business Review*, July 7, 2020.

109 Names can be a fraught issue: Jennifer Brown, host, "Minda Harts' 'The Memo': A Critical Alternative to Leaning In for Women of Color," *The Will to Change*, December 13, 2019.

111 The volume of job searches: "Coronavirus: How the World of Work May Change Forever," BBC, October 23, 2010, https://www.bbc.com/worklife/article/20201023-coronavirus-how-will-the-pandemic-change-the-way-we-work.

111 Bermuda and Barbados: Cailey Rizzo, "Bermuda Welcoming Remote Workers to Log On from Paradise with One-Year Residency Program," *Travel + Leisure*, July 23, 2020.

119 If you are concerned about potential ageism, Kamara Toffolo, a résumé and job search strategist: Andrew Seaman, "How to Overcome Common Resume and CV Stumbling Blocks," *LinkedIn News*, August 10, 2020.

128 Here are some examples: The final two headlines are adapted from Jon Shields, "12 Impactful LinkedIn Headline Examples from Real People," Jobscan blog, January 13, 2020.

131 In the first few months that #OpentoWork was available: Ryan Roslansky, "A New Look and Feel for LinkedIn," LinkedIn official blog, September 24, 2020.

Chapter 4: Networking in the "New Normal"

144 According to a 2016 report from the U.S. Bureau of Labor Statistics: Paige Harden, "How to Land a Job by Networking," *Washington Post*, May 23, 2016.

144 a 2018 study from HR Technologist found that referred talent: "3 Reasons Why Referrals Are the Way Forward for Recruitment in 2018," *HR Technologist*, January 26, 2018.

148 A Gallup study found that networking: John Clark, "College Alumni See Room for Job-Skill Improvement," Gallup, July 14, 2020.

148 searches for "how can I help?": Alyssa Fowers, "Last Year, We Searched Google for How to Tie a Tie. Now We're Using It to Find Toilet Paper," *Washington Post*, April 17, 2020.

151 The evolutionary psychologist Robin Dunbar: Mona Chalabi, "How Many People Can You Remember?" *FiveThirtyEight*, September 23, 2015.

159 Author and digital brand strategist Peter Thomson says: Peter Thomson, "How 50 Cups of Coffee Can Change Your Life," *Inc.*, September 13, 2013.

171 Recalculating has become so common: Jennifer Liu, "Nearly Half of Workers Have Made a Dramatic Career Switch, and This Is the Average Age They Do It," CNBC, *Make It*, October 31, 2019.

Chapter 5: Ace the Job Search

193 In a discussion titled: "A Guide to Surviving the Recession for New Grads," *Wall Street Oasis*, May 16, 2020.

197 According to one study, 75 percent of résumés: Kerri Anne Renzulli, "75% of Resumes Are Never Read by a Human—Here's How to Make Sure Your Resume Beats the Bots," CNBC, *Make It*, March 14, 2019.

197 "Neanderthal" of the five hundred résumé submissions writes that: "A Guide to Surviving the Recession for New Grads," *Wall Street Oasis*, May 16, 2020.

200 But Twitter's own cofounder: Connie Loizos, "Ev Williams: Twitter Should Be a Platform Company," *TechCrunch*, July 14, 2015.

205 The average life span of a company: Michael Sheetz, "Technology

Killing Off Corporate America: Average Life Span of Companies Under 20 Years," CNBC, August 24, 2017.

208 One study found that 64 percent of millennials: Morley Winograd and Michael Hais, "How Millennials Could Upend Wall Street and Corporate America," *Governance Studies at Brookings*, May 2014.

217 Even the U.S. Army: Arielle Wysocki, "US Army to Hold Virtual Hiring Campaign June 30–July 2," 104.5 WOKV, June 30, 2020.

220 According to Ian Siegel, cofounder and CEO of ZipRecruiter: Kathryn Dill, "The New Rules for Landing a Job in the Covid Era," *Wall Street Journal,* September 2, 2020.

224 Alexandra Carter, director of the Mediation Clinic at Columbia Law School: Kathryn Dill, "How to Ask for a Raise During the Covid Pandemic," *Wall Street Journal,* September 19, 2020.

233 Fortunately, this question is illegal: Jathan Janove, "More Jurisdictions Are Banning Salary-History Inquiries," Society for Human Resource Management, April 4, 2019.

233 In this case, Alexandra Carter recommends saying something like this: Dill, "How to Ask for a Raise During the Covid Pandemic."

234 Alexandra Carter recommends countering any hesitation or resistance: Ibid.

234 Salary is just one component of your total package: "Employer Costs for Employee Compensation Summary," U.S. Bureau of Labor Statistics, June 18, 2020.

Chapter 6: Turn Any Job into a Great Job

240 As one class of 2008 college grad shared: Megan Burbank, "How to Find a Job amid a Recession . . . or a Pandemic? Millennials Have This Advice for the Class of 2020," *Seattle Times*, May 28, 2020.

240 Richard Branson, the founder of Virgin Group: Richard Branson, "Want to Be More Productive? Be More Punctual," LinkedIn, October 5, 2015.

244 In June 2020, *Harvard Business Review* featured the story of Peter: Nihar Chhaya, "The Upside of Career Envy," *Harvard Business Review*, June 16, 2020.

246 Laura Vanderkam, the time-management expert, calls this "thanking people twice": Laura Vanderkam, host, "Thank People Twice," *Before Breakfast with Laura Vanderkam*, March 11, 2020.

261 In a 2020 survey from Robert Half, 78 percent of respondents with kids: "Survey: A Window into Windowed Work," Robert Half, May 14–19, 2020.

Chapter 7: Move On and Move Up

271 "If you say no to opportunities": Aliza Licht, "This Is How You Know You're Not Meant to Be an Entrepreneur," *Forbes*, November 24, 2016.

272 "[After graduating college into the global financial crisis in 2008], I really just enjoyed": This story was lightly adapted from Burbank, "How to Find a Job amid a Recession . . . or a Pandemic?"

280 "Don't be afraid to take a low-paying job": This story was lightly adapted from Burbank, "How to Find a Job amid a Recession . . . or a Pandemic?"

Conclusion

283 To quote my favorite tweet of 2020: Simon Holland (@simonholland), "Don't know about y'all but I could really go for some precedented times," Twitter, August 18, 2020, 7:05 A.M., https://twitter.com/simoncholland/status/1295678295296675840.

Index

About the Author

Lindsey Pollak is the *New York Times* bestselling author of *Becoming the Boss: New Rules for the Next Generation of Leaders*, *Getting from College to Career: Your Essential Guide to Succeeding in the Real World*, and *The Remix: How to Lead and Succeed in the Multigenerational Workplace*. She was named to the 2020 Thinkers50 Radar List of global management thinkers whose work is shaping the future of how organizations are managed and led. Her consulting and keynote speaking clients have included more than 250 corporations, law firms, conferences, and universities, including Aetna, Citigroup, Estée Lauder, GE, Google, and Stanford, and she has appeared in such media outlets as NBC's *Today*, the *New York Times*, the *Wall Street Journal*, CNN, and NPR. She is a graduate of Yale University and lives in New York City with her husband and daughter.